# Radiation Hormesis and the Linear-No-Threshold Assumption

Charles L. Sanders

# Radiation Hormesis and the Linear-No-Threshold Assumption

Springer

Charles L. Sanders, Ph.D.
Korea Advanced Institute of Science
and Technology (KAIST)
Department of Nuclear and Quantum
Engineering
335 Gwahak-ro (373-1 Guseong-dong)
Yuseong-gu, Daejeon 305–701
Republic of Korea

ISBN: 978-3-642-03719-1 e-ISBN: 978-3-642-03720-7

DOI: 10.1007/978-3-642-03720-7

Springer Heidelberg Dordrecht London New York

Library of Congress Control Number: 2009933986



*Cover design:* eStudio Calamar, Figueres/Berlin

Printed on acid-free paper

Springer is part of Springer Science+Business Media (www.springer.com)

# Preface

Outrageous, unsubstantiated statements are made concerning the hazards of ionizing radiation, in spite of a vast published, peer-reviewed literature on molecular, cellular, animal, and epidemiological studies indicating not harm but benefit from low-dose ionizing radiation. Claims that as many as a million children across Europe and Asia may have died in the womb as a result of radioactive fallout from Chernobyl or claims that the health impacts of low levels of internal radiation are underestimated by between 100 and 1,000 times are common among antinuclear arguments. Such statements are fueled by proponents of the linear nonthreshold (LNT) assumption, which assumes that any dose of radiation, no matter how insignificant, results in increased mortality from cancer and other diseases.

The most dishonest, manipulative research I have ever seen in my nearly 50 years of participation in radiobiological research has been published by radiation epidemiologists who are proponents of the LNT assumption. Their hundreds of publications and involvement in national and international radiation protection agencies have put them in a position of power and control within the research establishment. They have continued the deception in spite of overwhelming published, scientific data that clearly demonstrates how wrong the LNT assumption is. You might conclude that, if admirers of the LNT assumption were right about the risk from radiation, then the human race would not have survived natural background radiation, which in some places of the world is >100 mSv/year (>40 times the global average). The result of this deception is not insignificant: literally millions of lives are less healthy because they have been convinced that living in radiation deficient environments is healthy; lives are lost in not implementing effective low-dose radiation therapy to treat cancer; lives are lost out of fear of diagnostic radiation that saves lives; painful lives of people suffering from chronic inflammatory diseases are not improved by low-dose radiation therapy, which is given without the cost and side-effects of drugs and pain killers. Then there are the annual billions of dollars spent needlessly to protect us from radiation that we need for optimal health. Radiophobia limits the political will of people and governments to promote clean and safe nuclear power in place of traditional highly polluting fossil fuel power sources. Radiophobia prevents the logical and safe burial of nuclear wastes. Radiophobia causes serious psychological effects leading to loss of life (>100,000 abortions and >1,000 suicides attributed to Chernobyl fallout).

My research career was initially funded by AEC, starting at the Radiobiology Laboratory at Texas A & M under the leadership of Drs. Sidney Brown and George Krise from 1961 to 1963. All graduate students in the lab participated in a large reproduction study with rats who received continuous gamma irradiation from a $^{60}$Co source of ~20, 50, 100, or 200 mSv/day. I can remember their discussions about why rats receiving 20 mSv/day (~7 Sv per year) lived longer and had larger litter sizes and more robust reproduction than unexposed controls. In addition, they observed that tumor incidences in rats receiving 20 mSv per day were less than those in controls [1]. I spent about 25 years (1968–1993) working on inhalation toxicology of transuranics in the Biology department at PNL in Richland, WA. In the mid-1980s I became aware of the work of Dr. Luckey on radiation hormesis. My last project for DOE was concerned with establishing a dose–response relationship for inhaled high-fired $^{239}PuO_2$ in a study with 3,000 rats, including about 1,000 controls. An earlier, almost identical study published in 1976 examined the same issues in a smaller group of rats [2]. The new study included incorporation of tracer levels of $^{169}$Yb into plutonium oxide particles to facilitate accurate estimates of deep lung deposition of $^{239}$Pu. $^{169}$Yb delivered a lung dose of only 1–2 mSv from 175 keV γ-rays with the vast majority of the lung dose being delivered by α-particles from $^{239}$Pu. The result was a marked higher lung dose for first appearance of lung cancer, from 50 mGy for rats without $^{169}$Yb to 1,500 mGy for rats with $^{169}$Yb [2–4]. Thus, the start and end of my government funded research programs were punctuated by unexpected observations that strongly promoted the concept of radiation hormesis, which is the only believable explanation for these results.

This book will educate the interested general public and research and teaching professionals in the areas of radiobiology, health physics, medical physics, nuclear engineering, energy research, environmental science, medical and dental professions, and industry that low-dose radiation is not only safe but is healthy and beneficial.

Daejeon, Korea Charles L. Sanders

## References


1. Brown SO, Krise GM, Pace HB (1963) Continuous low-dose radiation effects on successive litters of the albino rat. Radiat Res 19:270–276
2. Sanders CL, Dagle GE, Cannon WC et al (1976) Inhalation carcinogenesis of high-fired 239PuO2 in rats. Radiat Res 68:340–360
3. Sanders CL, Scott BR (2008) Smoking and hormesis as confounding factors in radiation pulmonary carcinogenesis. Dose Response 6:53–79
4. Sanders C, Lundgren D (1995) Pulmonary carcinogenesis in the F344 and Wistar rat following inhalation of 239PuO2. Radiat Res 144:206–214

# Acknowledgments

I am grateful to the many people who made this book possible. Drs. Sidney Brown (deceased) and George Krise from the Radiobiology Laboratory, Texas A & M University, were wonderful professors who provided me with my first exposure to ionizing radiation as a research topic. Professors Gyuseong Cho, He Cheon No, Poong Hyun Seong, and Soon Heung Chang in the Department of Nuclear & Quantum Engineering at the Korea Advanced Institute of Science and Technology (KAIST) have been a great encouragement. Dr. Bobby Scott has unknowingly "mentored" me and continually kept me appraised of the latest developments. This book is built upon the pioneering work of Dr. Thomas Luckey, whose 1980 book on radiation hormesis left a profound impression. The subsequent work of Dr. Edward Calabrese continued and expanded the hormetic possibilities to toxic chemicals. Two KAIST graduate students, Sukwhun Sohn and Hosang Jeon, drew most of the illustrations. Five years of student classes listened to my passion and ramblings about the benefits of ionizing radiation. The Korean Nuclear Society, Korean Radiation Protection Society, KAIST and Seoul National University provided me the opportunity to freely disseminate my views throughout South Korea. The author acknowledges that the Lord made ionizing radiation to benefit his creation. *Everything is uncovered and laid bare before the eyes of him to whom we must give account* (Hebrews 4:13b).

# Contents

# Introduction 1

The use of the LNT is "faith-based radiation protection" [1]

Cancer arises from a variety of cell types with a prognosis that depends on tumor location and stage at time of diagnosis. The lifetime risk of fatal cancer in the U.S. is ~22% (23.6% for males and 19.9% for females) with lung, prostate, breast, and colorectal cancer being the most prominent [2]. In Korea, cancers of the stomach, breast, liver, and lung are the most prominent (Figs. 1.1 and 1.2). The average annual radiation dose to Americans and Koreans from natural sources (radon, internal radionuclides within the body, galactic–cosmic radiation, and primordial terrestrial sources, mostly from uranium and thorium) is about 2.5 mSv. The average annual dose from anthropogenic sources (mostly from medical sources) for both Americans and Koreans is about 0.5 mSv. The role of ionizing radiation

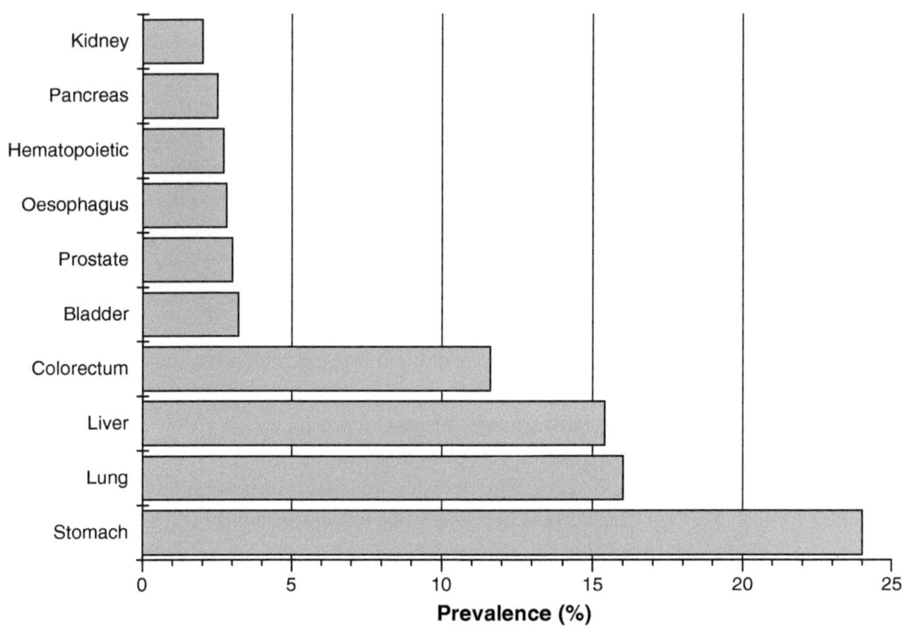


**Fig. 1.1** Prevalence of cancer in Korean men

C. L. L. Sanders, *Radiation Hormesis and the Linear-No-Threshold Assumption*,
DOI: 10.1007/978-3-642-03720-7_1, © Springer Verlag Berlin Heidelberg 2010

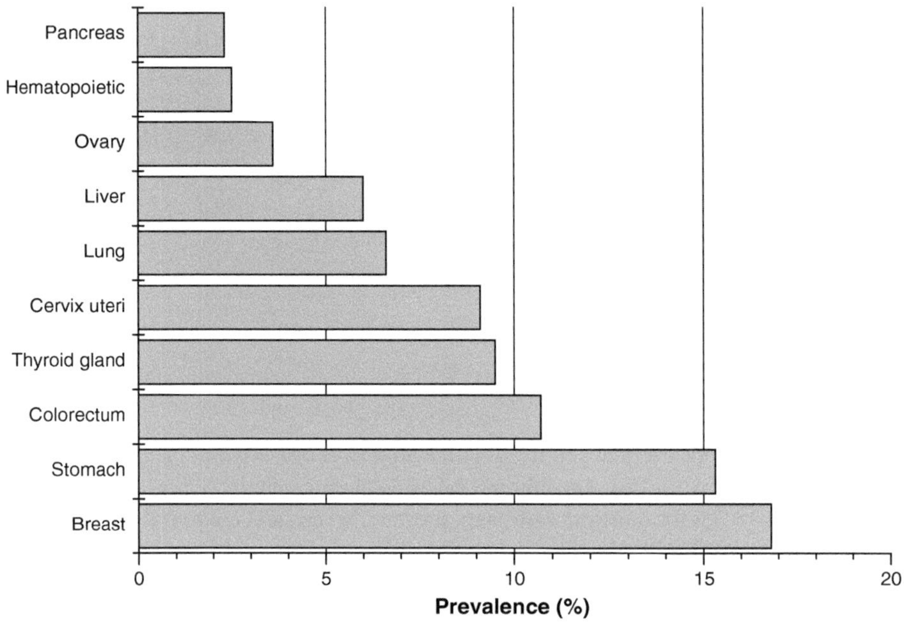


**Fig. 1.2** Prevalence of cancer in Korean women

in cancer formation at doses less than 1 Sievert (1 Sv = 1,000 mSv or 100 cSv) of low dose-rate, low LET (Linear Energy Transfer) radiation is the subject of this book.

The United States government is scheduled to spend $350 billion in cleaning up radioactive contamination and waste and decommissioning about 100 old nuclear power plants in 31 states during the next few decades. Current radiation protection standards established by the Environmental Protection Agency (EPA) were set using a linear extrapolation of World War II atomic bomb survivor data for cancer risk estimations. The standards are set for the general public at a small fraction above the natural background dose level, not taking into consideration the large variation in background dose levels throughout the world. Currently, anthropogenic radiation exposures to the general public are limited to 1 mSv year$^{-1}$. EPA nuclear cleanup exposure limit to the general public is 0.15 msv year$^{-1}$, while the Nuclear Regulatory Commission (NRC) uses a cleanup standard for decommissioning of nuclear power plants of 0.25 mSv [3].

Ionizing radiation is considered to be a weak carcinogen. Negative uncertainty about carcinogenic effects from ionizing radiation have influenced decommissioning of the existing nuclear facilities, long-term storage for reactor waste, construction and placement of new nuclear power plants, increased fears of "dirty bombs," and utilization of diagnostic radiology to find and treat disease. The decades-long moratorium on new construction of nuclear power plants in the U.S., a pervasive resentment of anything "nuclear," and a delay or refusal to obtain needed medical radiation exposures are some of the societal consequences to radiophobia among the American public [4, 5]. *While regulatory decision-making was designed to protect the public health, in some ways it has become punitive and burdensome. The idea that any exposure to radiation may be harmful has led to public anxiety and enormous economic expenditures that are disproportionate to the actual radiation risks involved. In the United States and some other countries, regulatory compliance costs are steadily growing, while desired public health benefits from added regulation are increasingly difficult to measure* [6]. A position paper of the Health Physics Society calls the regulatory systems for determining and enforcing public health standards "inconsistent, inefficient, and unnecessarily expensive" [7].

## 1.1 The LNT Assumption

In 1972, the Biological Effects of Ionizing Radiation (BEIR I) was published, using a linear model for risk estimates. Also in 1972, the United Nations Scientific Committee on Effects of Atomic Radiation issued UNSCEAR V, which questioned the validity of the linear model for radiation risk estimates. The LNT (Linear Non-Threshold) assumption is now widely accepted and applied, even though it has not been validated by scientific study and is not consistent with radiobiological data [8–10]. BEIR VII, ICRP, EPA, and NCRP support the LNT assumption for estimation of cancer risk from exposure to ionizing radiation [11–14].

The LNT assumption assumes a linear relationship between DNA damage in the form of double-strand breaks (DSB), that each DSB will have the same probability of inducing a cell transformation, and that each transformed cell will have the same probability of developing into a cancer [15]. Thus, cancer is thought to result from DNA (mutagenic) damage to a single cell caused by a single radiation track [16]. A low LET dose of 1 mGy is delivered to one cell nucleus by one electron track [14].

The LNT assumption is easy to implement utilizing the equivalent dose (biological damage weighted measure) and the effective dose (equivalent dose multiplied by a tissue-specific relative sensitivity factor for stochastic effects). The weighted doses are expressed in sievert (Sv) and millisievert (mSv, one-thousandth of a Sv). Expected cancer cases are easily calculated based on the summed effective dose (person-sievert) for an irradiated population [17]. The LNT assumption does not consider the role of biological defense mechanisms, but assumes that cancer risk proceeds in a proportionate linear fashion without a threshold to a point of zero dose through the origin. The LNT assumption with a low dose and dose rate effectiveness factor (DDREF) guarantees that any radiation dose, no matter how small, increases the risk of cancer. Lewis in 1951 was one of the first to determine the number of leukemia cases in the U.S., which could be attributed to background radiation by using the LNT assumption [18].

The use of the LNT assumption for purposes of radiation protection is assumed to be a cautious approach when applied to risk decision-making for human protection [19]. BEIR VI and BEIR VII [11], ICRP [19], and many epidemiologists and health physicists support the LNT assumption [20], while a large number of experimental and epidemiological studies challenge the validity of the LNT assumption, strongly suggesting the presence of a threshold and/or benefits from low doses of ionizing radiation [21–25].

The U.S. National Academy of Science supports the LNT assumption as a risk model of radiation-induced cancers. This means that the smallest dose of radiation causes cancer or

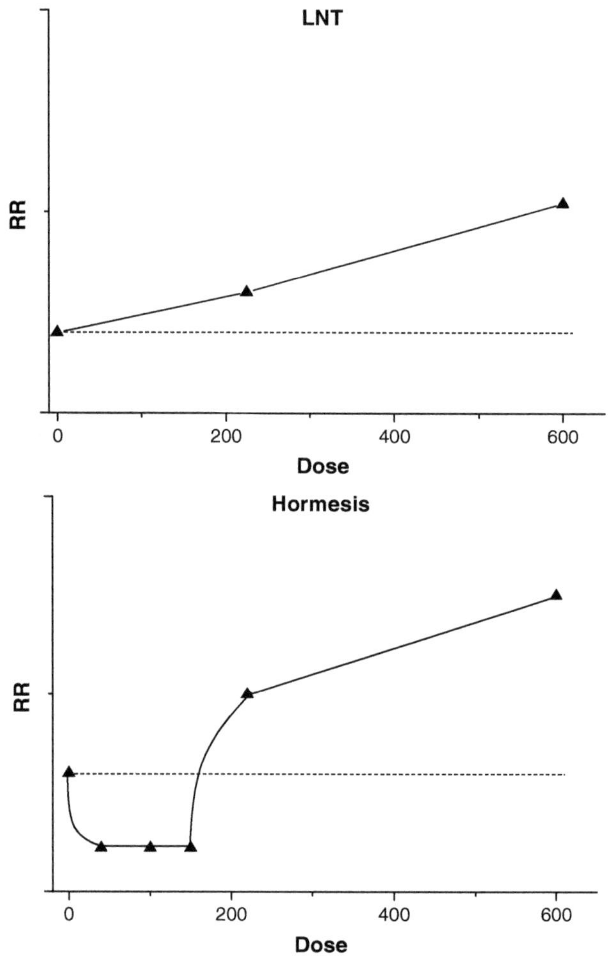


**Fig. 1.3** Graphic depiction of the LNT assumption and hormesis models

other health risks in humans [11]. As a result, cancer risk is a simple proportionality with dose, irrespective of dose-rate or LET (Fig. 1.3). This also implies that the mechanisms of cancer formation due to radiation are the same at low and high doses of ionizing radiation [21].

The LNT assumption is modeled mostly from epidemiological data of human populations exposed to high doses of ionizing radiation, but is assumed to apply to low doses and low dose-rates seen in occupational and accidental exposures. The LNT assumption was developed partly as a result of estimated effects from acute, high-dose atomic radiation attributed to nuclear weapons [26]. The high-dose effect predictions were easily extrapolated to low doses by assuming that any dose would contribute to the disastrous effects of nuclear war [27]. The LNT predicts:

1. Risk is linearly proportional to dose.
2. Every dose, no matter how small, carries a predictable risk.
3. There is no threshold.

4. The risk per unit of dose is constant, often expressed as excess relative risk (ERR).
5. Risk is additive.
6. Risk can only increase with dose.
7. Biological variables are insignificant when compared with dose.

Current international, radiological protection methods are based on the recommendations of the ICRP who utilize collective dose and the LNT assumption [19]. UNSCEAR estimates that the age- and gender-averaged risks of fatal solid cancer and leukemia following a whole-body acute dose of 1 Sv are 9.9 and 1.2%, respectively [4, 5]. Using the ICRP annual public dose limit of 1 mSv, a population of a million people receiving a dose of 1 mSv will have an expected number of excess cancers of $5 \times 10^{-5}$ per person mSv (derived from NCRP Report No. 115) $\times 1 \times 10^{6}$ person mSv, giving a collective dose of 50 Sv. The collective dose is then used to estimate total excess cancer deaths using simple linear extrapolation. Collective doses (for example 2,330,000 man Sv year$^{-1}$ from X-ray medical examinations [4, 5]) are meaningless results by multiplying tiny individual doses by 5.8 billion people. An example of the unfounded use of the LNT assumption is seen by reducing the natural background radiation dose from 0.05 to 0.0000000005 mSv, which would be expected to reduce cancer risk by a factor of 100,000,000 [1]. To emphasize the absurdity of such estimates, a collective dose of 14,000,000 man Sv year$^{-1}$ from natural sources was not given for comparison [27].

Computed Tomography (CT) scans deliver a radiation dose of 10–20 mSv. It is estimated that since 1980, more than 550 million CT scans have been obtained in the United States, 75 million of them before 1990 [28]. Using high dose data and the LNT assumption, Brenner and Hall estimate that 1.5–2% of all cancers in the United States are attributable to clinical use of CT [28].

Lauriston Taylor, one of the founders of the ICRP (International Commission of Radiological Protection) and NCRP (National Commission of Radiological Protection), wrote in 1980: *No one has been identifiably injured by radiation while working within the first numerical standards set first by the NCRP and then the ICRP in 1934* [29]. The safe limit for exposure in 1934 was ~0.2 rad day$^{-1}$, changing to 0.3 rad week$^{-1}$ in 1951, based on the concept of a threshold. By 1955, the threshold concept was rejected by ICRP. Under the new paradigm, excess cancers among radiologists and A-bomb survivors exposed to high doses are assumed to be stochastic with a probability of occurrence (not severity) being assumed to be proportional to dose.

The LNT controversy is being carried out in scientific (mechanistic) and policy (political) arenas. The validity of the LNT assumption has been challenged by many scientists [9, 10, 21, 30–33]. Abelson, editor of the journal, Science, criticized the LNT assumption in 1994: *To calculate effects of small doses, a linear extrapolation from large doses to zero is employed. The routine use of this procedure implies that the pathways of metabolism of large doses and small doses are identical. It implies that mammals have no defense against effects that injure DNA. It implies that no dose, however small, is safe. Examples of instances in which these assumptions are invalid are becoming numerous…The use of linear extrapolation from huge doses to zero implies that "one molecule can cause cancer." This assertion disregards the fact of natural large-scale repair of damaged DNA* [34].

The somatic mutation theory predicts that cancer begins from a single somatic cell mutation followed by successive mutations and other chromosomal/genetic changes [35].

Proponents believe that cancers are monoclonal. That is, tumors develop from the offspring of a single genetically damaged cell by a single radiation hit. A large practical threshold of 2–10 Gy is seen in humans for thorotrast patients (liver cancer) and radium dial painters (bone cancer) [36–38]. This contradicts the concept that a single a-particle will induce a cancer or even cause a cell transformation in vitro [39].

According to the LNT assumption, an increase in radiation dose increases the probability that a single cell will develop into a cancer. Thus, at low doses, a linearity of response is almost certain. BEIR VII assumes a linear relationship between low-LET dose and chromosomal mutations. Error-prone repair of double-strand DNA breaks, induced by a single ionizing cluster, is postulated as the important step in cellular neoplastic transformation leading to cancers [11].

The biological model of a single ionization event causing chromosomal damage to DNA in a cell resulting in a single mutation that produces a linear increase in cancer is not supported by research data. Paradoxical studies disprove the somatic mutation theory [40, 41]. Among the observations are widely distributed precancerous lesions, hyperplastic polyp genetic instabilities, spontaneous regression, a lower incidence of solid cancers in Down's syndrome, and a lack of tumors when carcinogen exposed epithelial cells are transplanted next to normal stroma [42]. Radiation-induced genomic instability (mutations, chromosome aberrations, cell death) appear in early stages of carcinogenesis, both *in vitro* and *in vivo*. These are frequent mutational events, consistent with a high frequency of transformed somatic cells [43]. Yet, the formation of a malignant tumor is exceedingly rare.

The LNT assumption is not supported by low LET data at acute doses <100 mSv or at chronic dose rates <200 mSv year$^{-1}$ [44]. Low doses and dose-rates of X- and $\gamma$-radiation upregulate various physiological mechanisms of protection against induction, propagation, and accumulation of cellular damage in tissues that may evolve into cancer [9, 45, 46]. The DOE Low Dose Radiation Research Program (1998–2008) has funded basic research to examine mechanistic responses at doses <100 mSv [47]. A recent discussion concerning the LNT model and the hormesis model has been published in Radiology [15, 16].

## 1.2 Radiation Hormesis and Radioadaptive Response

Hormesis is derived from the Greek word, *hormaein*, to excite. The concept of hormesis has its origins in the nineteenth century studies on the cellular pathology studies of Virchow [48]. Calabrese has provided several excellent historical reviews of hormesis [8, 49–51]. The hormesis hypothesis states that low-level stress (e.g., ionizing radiation) stimulates a system of protective biological processes at the cellular, molecular, and organismic levels, decreasing cancer incidence and the incidences of other deleterious health effects below spontaneous levels.

Hormesis is a dose–response phenomenon characterized by low dose stimulation and high dose inhibition. Mild stress-induced hormesis modulates and prevents aging and aging-related impairments. Low-level ionizing radiation is stimulatory at cellular, molecular, and organ levels. This radioadaptive response to low-dose radiation includes enhancement of antioxidant defenses, enzymatic repair of DNA, removal of DNA lesions, apoptosis, and immunologic stimulation, and is well established in the scientific literature [15, 52–60]. The benefits are inducible and transient, while the harmful stochastic effects of higher doses are seen after a long latency period. The effectiveness of these defense mechanisms varies with dose and dose-rate.

The radioadaptive response has been extensively studied, and is associated with increased lifespan as well as decreased mutations, chromosome aberrations, neoplastic transformations, congenital malformations, and cancer [61]. Apoptosis of chemically transformed, genomically damaged cells, including tumor cells, is activated by low-dose, low-LET $\gamma$ or x-radiations [62, 63]. Apoptosis removes cigarette-smoke and high LET radiation-induced, genomically damaged pulmonary cells before they can develop into lung cancers [23]. The protective adaptive response against cancer results from the traversal of only a single charged-particle track through the cell nucleus at very low dose rates. This protection against cancer risk may be as high as 75% at cumulative doses up to 20 mSv [61, 64, 65].

Low doses and dose rates of low-LET radiation protect mammalian cells *in vitro*, animals in experimental laboratory studies, and humans in clinical and epidemiological studies of radiation effects. These include protection against spontaneous genomic damage; protection against spontaneous and high-dose induced mutations; protection against spontaneous neoplastic transformation; protection against high dose chemical and $\alpha$-particle induced cancers; enhancement of immune system defenses against cancer; suppression of cancer metastases; protection against many noncancer diseases; and protection against heritable mutations and birth defects [66].

Low doses of ionizing radiation, like many toxic chemical compounds, improves health at low doses [67]. Broadly based data demonstrating hormesis has been found [68, 69]. Hormesis has been demonstrated for many diseases, including cardiovascular disease, diabetes, and cancer [8–10, 29]. Alcohol in the past was thought by most as evil and harmful at any dose. Today, many believe that alcohol is medicinal in low doses. Vitamin A in low doses prevents night blindness and pneumonia. But vitamin A in large doses is a deadly hepatotoxin. Large doses of radiation are harmful. Small doses make people healthier with less disease mortality than their under-exposed neighbors.

The U.S. National Cancer Institute (NCI) carried out cell proliferation screening studies for 2,189 possible chemical anti-tumor drugs in 13 different yeast strains. Retrospective evaluation of 57,000 dose–response studies showed a hormesis response pattern four times more often than expected [70]. Hormetic responses were found in 48% bioassays involving mice and in 14% bioassays involving rats in the National Toxicology Program (NTP) bioassay database, which included such toxic compounds as dioxin and methylcholanthrene [71]. Similar hormetic responses also exist for ionizing radiation.

In his book, *Radiation Hormesis* [52], Luckey describes evidences of radiation hormesis in workers at nuclear facilities, A-bomb survivors, and many other groups exposed to low doses of radiation. Luckey predicted that about one-third of all cancer deaths are preventable by low-dose ionizing radiation [15, 53, 54]. Calabrese has found about 3,000 examples of hormesis in the scientific literature [49–51].

A comprehensive research and development program on radiation hormesis was started in Japan in the late-1980s, and continues to provide very remarkable results. Excellent research on this subject was begun also in China at Jilin University, with special focus on

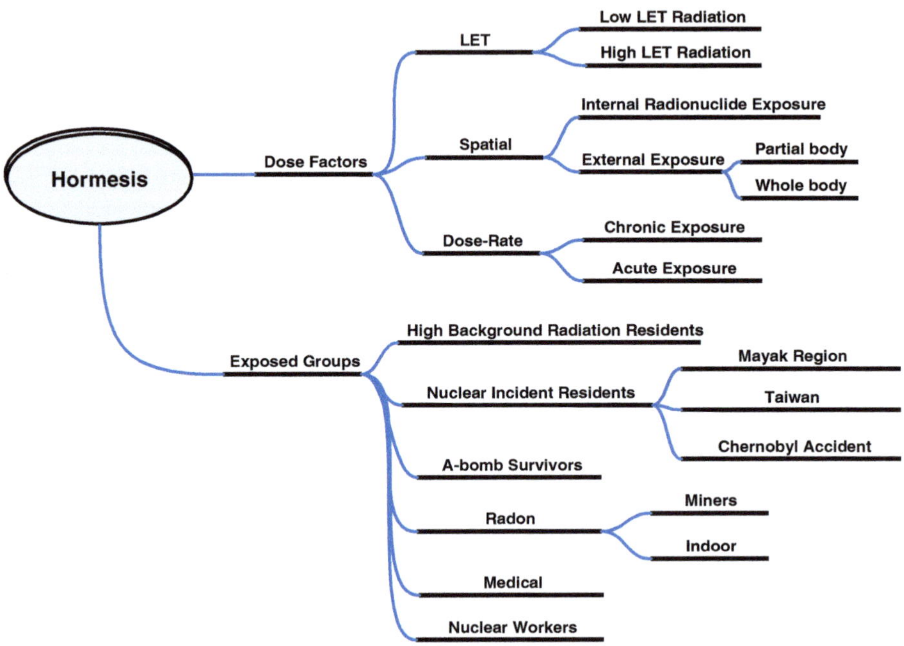


**Fig. 1.4** Factors associated with radiation hormesis

immune system stimulation with low dose radiation, and continues today in clinical cancer trials. In the late-1990s, the U.S. Department of Energy started an R&D program on low dose effects, which is producing evidence of an "adaptive response." Similar research is underway in Europe, the Middle East, and India.

No epidemiological data have demonstrated a significantly increased cancer rate in humans exposed to doses of ionizing radiation <100 mSv [21, 72]. In fact, many studies in a variety of exposed populations have demonstrated radiation hormesis (Figs. 1.4 and 1.5). Estimates of cancer risk in populations receiving cumulative radiation doses of <200 mSv cannot be made without using LNT extrapolation [73–76]. The 2005 French Academy of Sciences and National Academy of Medicine report concluded that the LNT assumption should not be used for low-LET doses <100 mGy [21]. The French Academies found abundant evidence for radiation hormesis and believed that this data should be implemented in making radiation protection guidelines.

Little thought has been given by radiation protection groups to radiation hormesis associated adaptive and inducible repair processes and thresholds at low doses and low dose-rates [77, 78]. *Opponents of the concept of hormesis, making use of strong appeals to authority, were successful in their misrepresentation of the scientific foundations of hormesis and in their unfair association of it with segments of the homeopathic movement with extreme and discreditable views. These misrepresentations became established and integrated within the pharmacology and toxicology communities as a result of their origins in and continuities with traditional medicine and subsequently profoundly impacted a broad range of governmental risk assessment activities further consolidating the rejection of hormesis. This error*

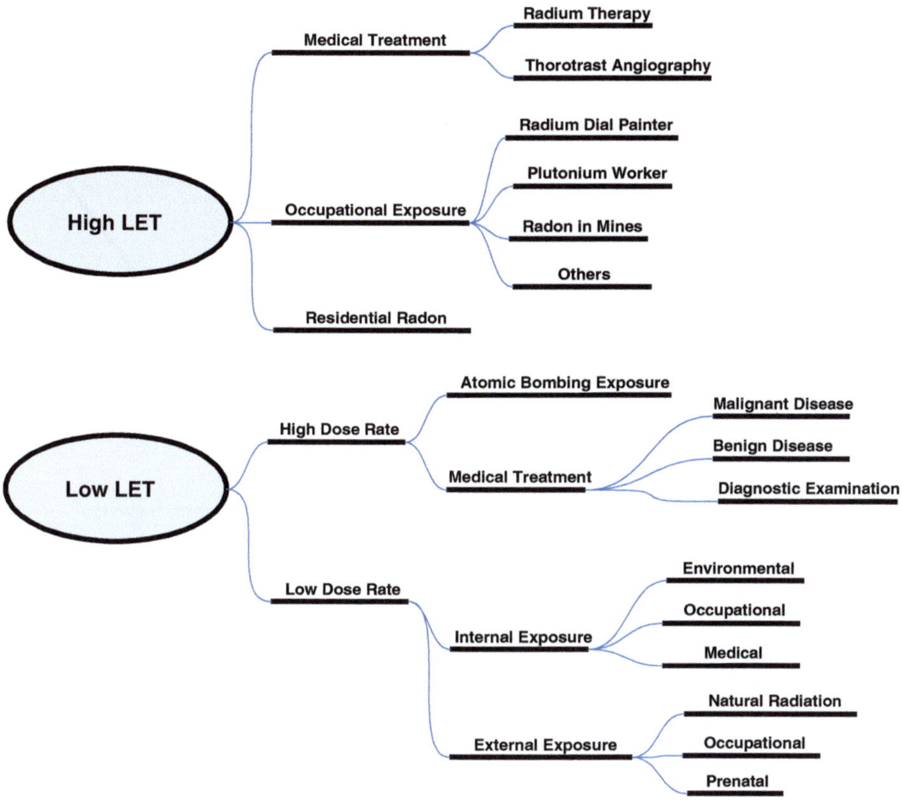


**Fig. 1.5** Exposure groups that demonstrate radiation hormesis

*of judgment was reinforced by toxicological hazard assessment methods using only high and few doses that were unable to assess hormetic responses, statistical modeling processes that were constrained to deny the possibility of hormetic dose–response relationships* [79].

Most experimental carcinogenesis data support a quadratic relationship with evidence of a practical or clearly defined threshold. The sigmoidal dose–response model birthed the concept of a threshold, which was the standard method used in toxicology up until 1954 when the NCRP replaced the concept of tolerance or threshold dose with permissible dose, suggesting that there was no "safe" dose of radiation [79]. The hormetic dose–response may be an inverted U-shaped or a J-shaped curve; the latter is often observed in disease incidence (Fig. 1.1). In many studies, the carcinogenic response to irradiation is suppressed at low doses and increased from a threshold dose in a stochastic manner at higher doses, leading to a curvilinear or U-shaped dose–response curve [9, 80].

According to the threshold model, there is a dose, below which there is no harm or there is a benefit. The threshold dose–response model is widely accepted in toxicology [81]. The threshold dose is approximated as a NOAEL (No Observed Adverse Effect Level). Hormesis with a threshold is operational even for those with genetic predispositions to

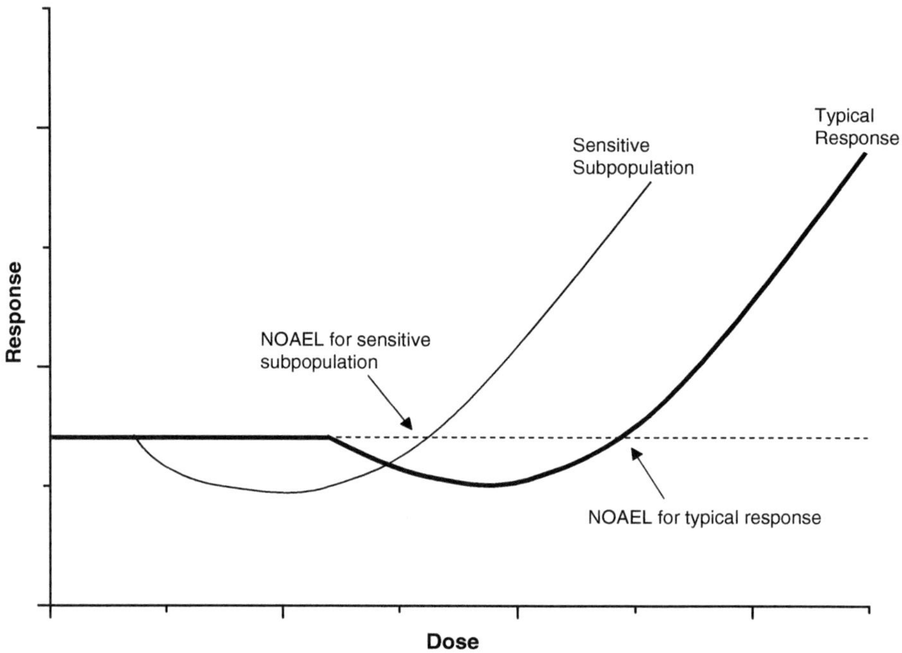


**Fig. 1.6** Hormetic responses in a typical normal population and in a genetically sensitive subpopulation [82]

cancer formation (Fig. 1.6). The controversy about hormesis is most intense with respect to estimating cancer risk, where the LNT assumption has dominated radiation protection agencies using high-to-low-dose extrapolations. On the other hand, application of the hormesis model will "optimize" health, not just protect against risk.

Some believe that a hormesis response, such as 50% less cancer than in control referent populations receiving less radiation doses, is modest [82]. Decreasing overall cancer mortality by 50% seems hardly insignificant, particularly at radiation doses that are below the NOAEL. In comparison, most estimates of radiation-induced cancer from low doses of radiation in epidemiological studies of nuclear, industrial, and medical workers using the LNT assumption give SMR (Standardized Mortality Ratio) or RR (Relative Risk) values <2.0, many at values <1.5, while values as low as 0.2 are seen for the radiation hormesis model at cumulative doses of <200 mSv.

The hormetic zone is a low dose region with the lower end usually including background radiation. A system of protective processes are considered to be maximally stimulated for low doses and dose rates of low LET radiation that are associated with the hormetic zone. For this dose zone, the relative risk at dose D (absorbed dose of all radiation involved) is given by the following:

$$\mathrm{RR} = 1, \text{ for } \mathrm{D} = 0,$$

$$\mathrm{RR} = 1 - \mathrm{PROFAC}, \tag{1}$$

The protection factor (PROFAC) gives the proportion of cancer cases (incidence or mortality) that are avoided among those cases that would have otherwise occurred spontaneously or in the absence of radiation hormesis. PROFAC, however, relates only to the low-LET component of any dose. Thus, for radon exposures, PROFAC relates to the gamma-ray component of the radiation dose. In the case of uranium mine radiation exposures, $\gamma$-rays may deliver from 25 to 75% of the "effective dose" to the lung [25]. High-LET $\alpha$-radiation by itself does not appear to activate the system of protective processes associated with radiation hormesis [83].

For exposure to low doses and dose rates of low-LET radiation or combinations of low- and high-LET radiations, rather than the dose–response curve increasing according to the LNT assumption, it instead is predicted to drop down to a constant value given by 1-PROFAC. There, however, are dose-rate-dependent transition zones (called transition zones A and B), where the individual-specific threshold for activating the protective hormesis process occurs (Fig. 1.7) [83]. Over this zone, the dose–response curve is expected to progressively decrease below RR = 1 rather than suddenly drop to 1 − PROFAC. In any case, the lowest point of the dose–response curve can be quantified as 1 − PROFAC. Further, this characterization of RR is expected to apply for most of the hormetic zone. Thus, where data are available for RR < 1 after low-dose, low-dose rates exposures to low or to low plus high-LET radiations, PROFAC can be justifiably estimated using:

$$\text{PROFAC} = 1 - \text{RR}. \tag{2}$$

$$\text{ERR} = \text{RR} - 1 \tag{3}$$

Systematic errors are expected to lead to underestimation of the PROFAC when the equation is used for moderate and high doses with a more complicated equation where hormesis can be suppressed. For doses in Transition Zone A, systematic error is also expected to favor underestimation of PROFAC. However, for a large portion of the hormetic zone, RR = 1-PROFAC is expected to adequately characterize the dose–response relationship [84]. The product 100•PROFAC allows representing the cancer cases avoided as a percentage (%) rather than as a proportion.

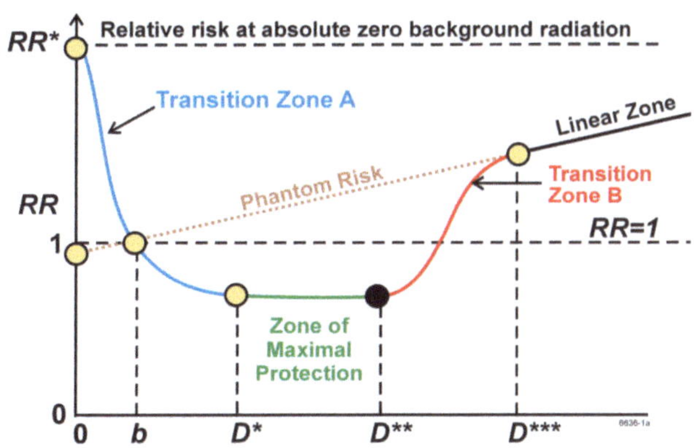


**Fig. 1.7** Taken from Scott (2005) [83]

Risk is often expressed in this book as relative risk (RR). Also commonly used is the Standardized Mortality Ratio (SMR). The SMR can be used as an estimate of RR allowing PROFAC to be obtained based on SMR data. An Odds Ratio (OR) $<1$ can also serve as an estimate of RR. For rare (low probability) diseases, OR estimates RR in a reliable manner. Irradiated persons should be excluded as much as possible in unexposed groups. Otherwise, large systematic error can occur leading to changing a threshold-type hormetic curve into what appears to be an LNT curve. RR, SMR, and OR data have been used to estimate PROFAC for irradiated human populations where hormetic effects have been demonstrated or suspected.

Values for PROFAC significantly $>0$ implicate spontaneous cancers being avoided that would otherwise have occurred (possible fatal) in the absence of radiation hormesis. It should be kept in mind that what are considered spontaneous cancers may actually be associated with other environment-associated risk factors. However, radiation hormesis would also be expected to protect against stochastic effects caused by such factors.

For a population residing in a high natural background radiation area where a significant component of the radiation dose is due to low-LET radiation, for each 100,000 spontaneous cancer deaths that would be expected over a given follow-up period in the absence of radiation hormesis, each increment of 0.1 in the PROFAC (due to radiation hormesis) would be expected to save 10,000 lives. Thus, radiation hormesis is not a trivial benefit. Since many reported PROFACs exceed 0.3, this translates into very large numbers of lives being saved due to radiation hormesis.

The PROFAC depends on dose rate, the type of radiation (i.e., radiation quality), and the target organ in the body. Equations 1 and 2 should be applied to low doses and dose rates only where the radiation exposure includes low-LET radiation. The same PROFAC is considered to apply to cancer incidence and cancer mortality [84].

Special dosimetric considerations are used to estimated radiation dose for high-LET radiation. The Working Level (WL), where 1 WL = 3,700 Bq $m^{-3}$ or 100 pCi $L^{-1}$ of $^{222}Rn$ in equilibrium with short-lived decay products, is used to determined radon exposure. One Working Level Month (WLM) corresponds to an exposure to one Working Level (WL) for 170 h [85]. WLM was converted to dose in mSv by the coefficient, 5.06 mSv per WLM, as recommended by ICRP [85]. Others have calculated a factor of up to 14 mSv per WLM [86, 87]. Lung dose from indoor radon was calculated using a coefficient of 25 μSv $year^{-1}$ per Bq $m^{-3}$, assuming an indoor occupancy of 0.8 [88]. The coefficient was multiplied by 57, the average lung cancer patient's lifespan, assuming an indoor occupancy of 0.8 [89]. For experimental animals exposed to pulmonary $\beta$ and $\alpha$ irradiation, an RBE of 10 is used to convert dose in Gy for $\alpha$-radiation to dose in Sv [90]; an RBE of 10 was used to determine dose in Sv for this book. The ICRP recommended value of 20 weighting factor for alpha particles for uranium miners appears to be to high [85, 91].

## References

1. Dr Bobby Scott, personal communication
2. Jemal A, Murray T, Samuels A et al (2003) Cancer Statistics, 2003. CA Cancer J Clin 53:5–26

3. General Accounting Office (2000) Radiation Standards. Scientific Basis Inconclusive, and EPA and NRC Disagreement Continues, Report to the Honorable Pete Domenici, US Senate, GAO/RCED-00–152, General Accounting Office
4. UNSCEAR (2000) Sources and effects of ionizing radiation, vol II. Effects. Annex II, United Nations, New York
5. Strzelczyk J, Potter W, Zdrojewicz Z (2007) Rad-by-rad (bit-by-bit): triumph of evidence over activities fostering fear of radiogenic cancers at low doses. Dose-Response 5:275–283
6. Mossman KL et al (2000) Final report, Bridging Radiation Policy and Science, An International Conference, Airlie House Conference Center, Warrenton, VA, pp 1–5
7. Health Physics Society (2001) Compatibility in radiation-safety regulations. Position paper of the Health Physics Society
8. Calabrese EJ, Baldwin LA (2000) Radiation hormesis: its historical foundations as a biological hypothesis. Hum Exper Toxicol 19:41–75
9. Aurengo A, Averbeck D, Bonnin A et al (2005) Dose-effect relationships and estimation of the carcinogenic effects of low doses of ionizing radiation. Executive Summary. French Academy of Sciences, French National Academy of Medicine
10. Tubiana M (2003) The carcinogenic effect of low doses: the validity of the linear no-threshold relationship. Intern J Low Radiation 1:1–33
11. Committee to Assess Health Risks from Exposure to Low Levels of Ionizing Radiation (2005) BEIR VII – Phase 2, Health Risks from Exposure to Low Levels of Ionizing Radiation, National Research Council (National Academy of Sciences)
12. ICRP Draft Report of Committee I/Task Group (2004) Low dose extrapolation of radiation related cancer risk, Dec
13. EPA (Environmental Protection Agency) (1994) Estimating radiogenic cancer risks. EPA Report 402-R-93–076, Washington, DC
14. NCRP (National Council on Radiation Protection and Measurements) 2001 Evaluation of the linear-non-threshold dose-response model for ionizing radiation. NCRP Report No. 136, Bethesda, MD
15. Tubiana M, Feinendegen LE, Yang C, Kaminski JM (2009) The linear no-threshold relationship is inconsistent with radiation biologic and experimental data. Radiology 251:13–22
16. Little MP, Wakeford R, Tawn EJ et al (2009) Risks associated with low doses and low dose rates of ionizing radiation: why linearity may be (almost) the best we can do. Radiology 251:6–12
17. Scott BR (2008) It's time for a new low-dose-radiation risk assessment paradigm—one that acknowledges hormesis. Dose-Response 6:333–351
18. Lewis E (1957) Leukemia and ionizing radiation. Science 43:965
19. International Commission on Radiological Protection (1991) Recommendations of the International Commission on Radiological Protection. Elsevier Science, New York. ICRP Publication No. 60, Ann ICRP 21:1–3
20. Gilbert ES (2009) Ionising radiation and cancer risks: What have we learned from epidemiology? Int J Radiat Biol 85:467–482
21. Tubiana M, Aurengo A, Averbeck D, Masse R (2006) The debate on the use of linear or threshold for assessing the effect of low doses. J Radiol Prot 26:317–324
22. Duport P (2003) A database of cancer induction by low dose radiation in mammals: overview and initial observations. Int J Low Radiat 1:120–131
23. Sanders CL, Scott BR (2008) Smoking and hormesis as confounding factors in radiation pulmonary carcinogenesis. Dose-Response 6:53–79
24. Tanooka H (2001) Threshold dose-response in radiation carcinogenesis: an approach from chronic beta-irradiation experiments and a review of non-tumour doses. Int J Radiat Biol 77:541–551
25. Duport P (2002) Is the radon risk overestimated? Neglected doses in the estimation of the risk of lung cancer in uranium underground miners. Radiat Prot Dosim 98:329–338

26. Glasstone S (1957) The effects of nuclear weapons. United States Department of Defense and United States Atomic Energy Commission, Washington, DC
27. Jaworowski Z (2009) Radiation hormesis—a remedy for fear. BELLE Newsletter 15:14–20
28. Brenner DJ, Doll R, Goodhead DT et al (2003) Cancer risks attributable to low doses of ionizing radiation: assessing what we really know. Proc Natl Acad Sci U S A 100: 13761–13766
29. Taylor LS (1980) Some non-scientific influences on radiation protection standards and practice. Health Phys 32:851–874
30. Bond VP, Wielopolski L, Shani G (1996) Current misinterpretation of the linear no threshold hypothesis. Health Phys 70:877–882
31. Latarjet R, Tubiana M (1989) The risks of induced carcinogenesis after irradiation at small doses. The uncertainties which remain after the 1988 UNSCEAR report. Int J Radiat Oncol 17:237–240
32. Rossi HH, Zaider M (1997) Radiogenic lung cancer. The effect of low doses of low LET radiation. Radiat Environ Biophys 36:85–88
33. Gros F (1999) Carcinogenic risks due to ionizing radiation. Life Sciences. CR Acad Sci [III] 322:81–256
34. Abelson PH (1994) Risk assessment of low level exposure. Science 265:1507
35. Michor F, Iwasa Y, Nowak M (2004) Dynamics of cancer progression. Nature Review Cancer 4:197–205
36. Carnes BA, Groer PG, Kotec TJ (1997) Radium dial workers: issues concerning dose response and modeling. Radiat Res 147:707–714
37. Nyberg U, Nilsson B, Travis LB et al (2002) Cancer incidence among Swedish patients exposed to radioactive thorotrast: a forty-year follow-up survey. Radiat Res 157:419–425
38. Van Kaick G, Dalheimer A, Hornik A et al (1999) The German thorotrast study: recent results and assessment of risks. Radiat Res 152:564–572
39. Miller RC, Randers-Pehrson G, Geand CR et al (1999) The oncogenic transforming potential of the passage of single alpha particles through mammalian cell nuclei. Proc Natl Acad Sci U S A 96:19–22
40. Sonnenschein C, Soto A (2000) Somatic mutation theory of carcinogenesis: why it should be dropped and replaced. Molecular Carcinogenesis 29:205–211
41. Soto A, Sonnenschein C (2004) The somatic mutation theory of cancer: growing problems with the paradigm. Bioessays 26:1097–1107
42. Baker SG, Kramer BS (2007) Paradoxes in carcinogenesis: new opportunities for research directions. BMC Cancer 7:151–156
43. Morgan WF (2003) Non-targeted and delayed effects of exposure to ionizing radiation: II. Radiation-induced genomic instability and bystander effects in vivo, clastogenic factors and trans-generational effects. Radiat Res 159:581–596
44. Higson DJ (2004) The bell should toll for the linear no-threshold model. J Radiol Prot 24:315–319
45. Feinendegen LE, Neumann RD (2005) Physics must join with biology in better assessing risk from low-dose irradiation. Radiat Prot Dosimetry 117:346–356
46. Feinendegen LE, Pollycove M, Neumann RD (2007) Whole-body responses to low-level radiation exposure: new concepts in mammalian radiobiology. Exp Hematol 35(Suppl 1):37–46
47. Council of Scientific Society Presidents (1998) Creating a strategy for science-based national policy: addressing conflicting views on the health risks of low-level ionizing radiation, Wingspread Conference, Racine, WI, July 31-August 3, 1997
48. Henschler D (2006) The origin of hormesis: historical background and driving forces. Hum Exp Toxicol 25:347–351
49. Calabrese EJ (2002) Hormesis: changing view of the dose response: a personal account of the history and current status. Mutat Res 511:181–189

50. Calabrese EJ, Baldwin LA (2001) Hormesis: a generalizable and unifying hypothesis. Crit Rev Toxicol 31:353–424
51. Calabrese EJ, Baldwin LA (2001) The frequency of U-shaped dose responses in the toxicological literature. Toxicol Sci 62:330–333
52. Luckey TD (1991) Radiation hormesis. CRC Press, Boca Raton, FL
53. Luckey TD (1982) Physiological benefits from low-level ionizing radiation. Health Phys 43:771–789
54. Luckey TD (1999) Nurture with ionizing radiation: a provocative hypothesis. Nutrition and Cancer 34:1–11
55. Kant K, Chauhan RP, Sharma GS et al (2003) Hormesis in humans exposed to low-level ionizing radiation. Intern J Low Radiat 1:76–87
56. Jaworowski Z (1995) Stimulating effects of ionizing radiation: new issues for regulatory policy. Regulatory Toxicology and Pharmacology 22:172–179
57. Calabrese EJ, Baldwin LA (2001) Scientific foundations of hormesis. Crit Rev Toxicol 31:351–624
58. Pollycove M, Feinendegen LE (2001) Biologic responses to low doses of ionizing radiation: detriment versus hormesis. Part 2: Dose responses to organisms. J Nucl Med 42:26N–37N
59. Pollycove M, Feinendegen LE (1999) Molecular biology, epidemiology and the demise of the linear no-threshold (LNT) hypothesis. In: CR Acad. Sci., Paris, Life Sciences 322:197–204
60. Pollycove M (1998) Nonlinearity of radiation health effects. Environ Health Perspect 106:363–368
61. Leonard BE (2007) Adaptive response and human benefit: Part I. A microdosimetry dose-dependent model. Int J Radiat Biol 83:115–131
62. Stephens LC, Kang K, Schultheiss TE et al (1991) Apoptosis in irradiated murine tumors. Radiat Res 127:308–316
63. Thompson HJ, Strange R, Schedin PJ (1992) Apoptosis in the genesis and prevention of cancer. Cancer Epidemiol Biomarkers Prev 1:597–602
64. Leonard BE (2005) Adaptive response by single cell radiation hits-implications for nuclear workers. Radiat Prot Dosimetry 116:387–391
65. Leonard BE (2007) Adaptive response: Part II. Modeling for dose rate and time influences. Int J Radiat Biol 83:395–408
66. Scott BR, SA Belinsky, S Leng et al (2009) Radiation-stimulated epigenetic reprogramming of adaptive-response genes in the lung: an evolutionary gift for mounting adaptive protection against lung cancer. Dose Response 7:104–131
67. Hiserodt E (2005) Under-exposed. What if radiation is actually good for you? Laissez Faire Books, Little Rock, AR, p 247
68. Calabrese EJ, Baldwin LA (2003) Hormesis: the dose-response revolution. Ann Rev Pharmacol Toxicol 43:175–197
69. Calabrese EJ (2005) Paradigm lost, paradigm found: The re-emergence of hormesis as a fundamental dose response model in the toxicological sciences. Environmental Pollution 138:378–411
70. Calabrese EJ, Staudenmayer JW, Stanek EJ et al (2006) Hormesis outperforms threshold model in National Cancer Institute antitumor drug screening database. Toxicol Sci 94: 368–378
71. Calabrese EJ, Baldwin LA (2003) Hormesis at the National Toxicology Program (NTP): evidence of hormetic dose responses in NTP dose-range studies. Nonlinearity in Biology, Toxicology, and Medicine 1:455–467
72. Kondo S (1993) Health effects of low-level radiation. Kinki University Press, Osaka, Japan and Medical Physics Publishing, Madison, WI
73. Berrington A, Darby S (2004) Risk of cancer from diagnostic X-rays: estimations for the UK and 14 other countries. Lancet 363:345–351

74. Breckow J (2006) Linear-no-threshold is a radiation-protection standard rather than a mechanistic effect model. Radiat Environ Biophys 44:257–260
75. Cardis E, Vrijheid M, Blettner M et al (2005) Risk of cancer after low doses of ionizing radiation: retrospective cohort study in 15 countries. BMJ 331:77–83
76. Darby S, Hill D, Auvinen A et al (2005) Radon in homes and risk of lung cancer: collaborative analysis of individual data from 13 European case-control studies. BMJ 330:223–228
77. Feinendegen LE, Bond VP, Booz J et al (1988) Biochemical and cellular mechanisms of low-dose effects. Int J Radiat Biol 53:23–37
78. Shadley JD, Wiencke JK (1989) Induction of the adaptive response by X-rays is dependent on radiation intensity. Int J Radiat Biol 56:107–118
79. Calabrese EJ (2005) Historical blunders: how toxicology got the dose-response relationship half right. Cellular and Molecular Biology 51:643–654
80. Calabrese EJ, Baldwin LA (2003) The hormetic dose-response model is more common than the threshold model in toxicology. Toxicol Sci 71:246–250
81. Eaton DL, Klassen CD (2001) Principles of toxicology. In: Klassen CD (ed) Casarett & Doull's toxicology: the basic science of poisons, 6th edn. McGraw-Hill, New York, pp 11–34
82. Hoffman GR and WE Stempsey. 2008. The hormesis concept and risk assessment: Are there unique ethical and policy considerations. BELLE 14(3):11–17
83. Scott BR (2005) Stochastic thresholds: a novel explanation of nonlinear dose-response relationships. Dose-Response 3:547–567
84. Scott BR (2007) Low-dose radiation-induced protective process and implications for risk assessment, cancer prevention, and cancer therapy. Dose-Response 5:131–141
85. ICRP (International Commission on Radiological Protection) (1994) Protection against radon-222 at home and at work. ICRP Publication 65. Ann ICRP 23, Pergamon, Oxford
86. Birchall A, James AC (1994) Uncertainty in analysis of the effective dose perunit exposure from 222Rn D.P. and implications for ICRP risk-weighting factors. Radiat Prot Dosim 53:133–140
87. Marsh JW, Birchall A, Butterweck G et al (2002) Uncertainty analysis of the weighted equivalent lung dose per unit exposure to radon progeny in the home. Radiat Prot Dosim 102:229–248
88. UNSCEAR (1993) Sources and effects of ionizing radiation. United Nations Scientific Committee on the Effects of Atomic Radiation (UNSCEAR), United Nations, New York
89. Steck DJ (1990) Indoor radon and lung cancer in Minnesota. In: Proceedings of the Technical Exchange Meeting on Assessing Indoor Radon Health Risks, September, 1989, Grand Junction, CO. U.S. Department of energy CONF-8909190
90. Sanders CL, Thompson RC, Bair WJ (1970) Lung cancer: dose response studies with radionuclides. In: Inhalation Carcinogenesis, CONF-691001, NTIS, Springfield, VA, pp 285–303
91. Rattan SIS, Fernandes RA, Demirovic D et al (2009) Heat stress and hormetin-induced hormesis in human cells: effects on aging, wound healing, angiogenesis, and differentiation. Dose-Response 7:90–103

# Molecular and Cellular Mechanisms 2

I have always felt that the argument that because at higher vaules of dose an observed effect is proportional to dose, then at very low doses there is necessarily some "effect" of dose, however small, is nonsense [1].

## 2.1 Introduction

The adult human body has more than 100 trillion cells comprising ~100 cell types. There are $10^{11}$ cell divisions a day, resulting in $10^{16}$ new cells during a 70-year lifespan. DNA in each cell is attacked by reactive oxygen species (ROS) a million times a day (Table 2.1) [2, 3]. Yet, the chance of one of these cells undergoing genetic changes leading to a lethal tumor in 70 years is about 20%. This means that there are powerful and highly effective protective mechanisms within human cells that limit cancer expression.

Ionizing radiation is a weak carcinogen even at high doses. As a result, the carcinogenic effect of radiation is difficult to statistically demonstrate in human populations, particularly at low doses [5]. The LNT model hypothesizes that energy deposition in the cell nucleus results in damage to DNA that is proportionately responsible for cancer formation. However, cancer incidences at low doses in irradiated human populations are not consistent with predictions using the LNT assumption [6].

Muller found that X-irradiated male fruit flies developed increased sex-linked mutations when crossed with nonirradiated female flies [7]. A subsequent study found a linear relationship between dose to fruit flies and mutation frequency over a dose range of 3–44 Gy (Fig. 2.1) [8]. The International Commission on Radiological Protection (ICRP) used these results on fruit fly mutations as the basis to estimate radiation-induced stochastic risks in humans [9], which later was applied as the LNT assumption. Even so, epidemiological studies have not detected statistically significant hereditary effects from ionizing radiation in humans [10]. UNSCEAR (2001) gave a doubling dose for genetic effects in humans of 3.4–4.5 Gy, based on estimates of genetic damage in Japanese A-bomb survivors, even though there were no genetic effects in A-bomb survivors [11].

The dose–response relationship in irradiated DNA repair-proficient fruit flies is not linear, showing a threshold at 1 Gy for somatic mutations [12]. In addition, the mutation

C. L. L. Sanders, *Radiation Hormesis and the Linear-No-Threshold Assumption*,
DOI: 10.1007/978-3-642-03720-7_2, © Springer Verlag Berlin Heidelberg 2010

**Table 2.1** Comparison of DNA damage from reactive oxygen species (ROS) and a background radiation level of 2 mSv year$^{-1}$

| Type | Spontaneous ROS | 2 mSv year$^{-1}$ |
|---|---|---|
| DNA oxidative adducts/ cell/day [2] | $10^6$ | $1 \times 10^{-3}$ |
| DNA Damage per cell per year [3] | $70 \times 10^6$ | 4 |
| Double-stranded DNA breaks per cell per year [4] | 40 | ~0.1 |

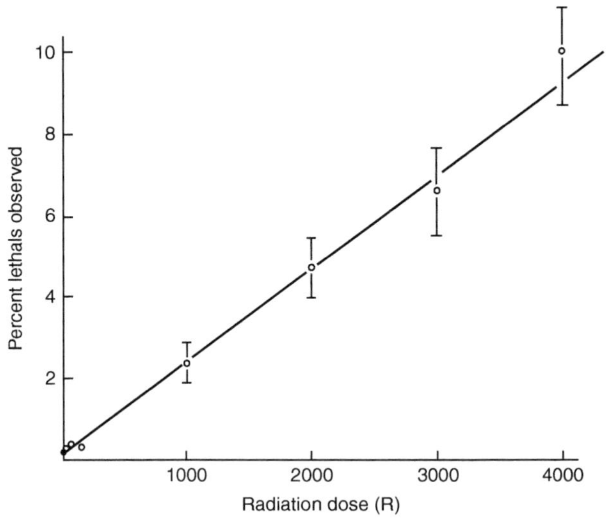


**Fig. 2.1** Mutation rate in X-irradiated *Drosophila* spermatozoa [8]

frequency in wild-type flies, where spermatozoa were irradiated at a total dose of 0.2 Gy and dose-rate of 0.05 Gy min$^{-1}$, was significantly lower than the mutation frequency found in sham-irradiated flies [13]. The mutation frequency of sex-linked recessive lethal mutations in *Drosophila melanogaster* irradiated with 500 μGy was significantly lower than in the control group ($p < 0.01$) [14]. Irradiation of mutant flies defective in DNA excision repair did not show a reduction in mutation frequency [15]. In summary, fruit fly mutation data does not provide any evidence for linearity at low doses and should not be used as apologetic evidence supporting the LNT assumption, since there is good evidence for the presence of a threshold and hormesis at low doses.

Gene activation and repression and gene transcription are different at low doses than at high doses [16]. Low-dose radiation activates genes, whereas high-dose radiation depresses the same set of genes [17, 18]. Low-dose-rate photon irradiation of mammalian cells causes a lower frequency of induced mutations than high-dose-rate irradiation [19]. HPRT mutation induction is one-eighth at a dose rate of 0.3 mGy min$^{-1}$ of that seen at a dose-rate of 2 Gy min$^{-1}$ [20]. Such qualitative and quantitative differences in gene responses at low and high doses further demonstrates the error in extrapolation of cancer risk from high to low doses [21, 144].

The average adult breathes 500 g oxygen per day. ROS causes a very large number of oxidant attacks per cell per day. Genetic damage to cells generated by normal cellular oxygen metabolism is more than a 1,000 times greater than damage induced by background radiation (Table 2.1) [22, 23]. DNA damage caused by ROS per minute is equivalent to damage caused by a dose-rate of a few mGy per minute. There is no detectible excess in mutations or cancer at this dose rate when the total dose is <200 mGy [24].

The concept of a negligible or safe dose is central to establishing exposure standards in radiological health. The radioadaptive protective response is thought to result from the traversal of a cell nucleus by only a single charged particle at very low-dose rates [25–27]. Low-LET radiation accounts for 50–85% of radiation damage under aerobic conditions from the indirect action of induced radicals on DNA. High-LET radiation produces direct damage to DNA that is less likely to be repaired [28]. Misrepaired double-stranded breaks (DSBs) in DNA may be the principle molecular lesions of importance in carcinogenesis [23].

## 2.2 The Radioadaptive Response

Low-dose radiation stimulates many cellular functions that protect DNA from damage. This radioadaptive response is well established in the scientific literature [29–37]. The benefits are inducible and transient. The carcinogenic response to irradiation increases faster in a stochastic manner at higher doses, leading to a nonlinear dose–response relationship, since the effectiveness of these defense mechanisms varies with dose and dose rate [51]. Stimulation at low doses is in contrast to damage at high doses precludes a linear relationship between radiation dose and health effects [7]. Epigenetic changes, such as hypomethylation of DNA, are linked to increased aging and mutation and cancer rates [38, 39]. Hypermethylation can also be interpreted as a radioadaptive response [40], which helps to explain the hormesis effect found in experimental and epidemiological studies [41, 42].

The adaptive response involves DNA repair processes, radical scavenging induction, apoptosis, cell transformation, and cell cycle arrest [43, 44]. There are three major cellular defense systems against ionizing radiation that comprise the radioadaptive response (Fig. 2.2) [21, 29, 30, 37, 45–47]: (1) protection against reactive oxygen species (ROS) by antioxidant molecules (such as glutathione) and detoxifying enzymes (such as catalase and superoxide dismutase); (2) DNA repair, particularly for double-strand breaks, that disappears at doses >0.5 Gy; and (3) elimination of genomically damaged cells by immune defenses and apoptosis at doses as low as a few mSv. The hormesis response is associated with increased lifespan, and decreased mutations, chromosome aberrations, neoplastic transformations, cancer, and congenital malformations [48].

The various parts of the adaptive response can be maintained for a few hours to several months [40]. Protection against ROS lasts a few hours, DNA repair lasts a few days, and immune enhancement may last up to several months (Fig. 2.3) [49].

The adaptive response occurs at dose values ranging from 0.01 to 0.5 Gy and at dose-rate values ranging from 0.01 to 1.0 Gy $min^{-1}$ [40]. The radioadaptive response appears most beneficial at doses <100 mGy [49]. The response begins to disappear at doses >200 mGv of low-LET radiation and is rarely seen at a dose of >500 mGv. However, the probability of apoptosis appears to increase in a linear fashion beyond 500 mGv [49].

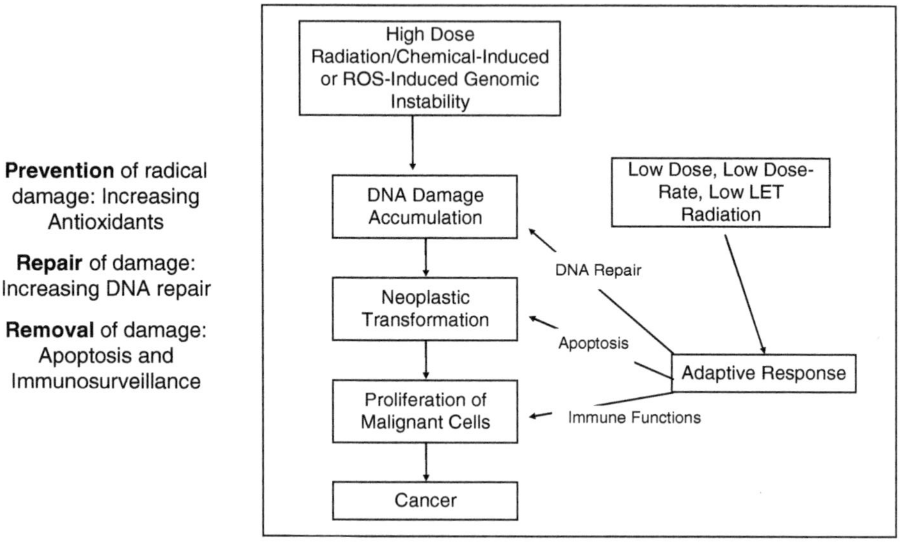


**Fig. 2.2** Mechanisms of prevention, repair, and removal of ROS and radiation damage

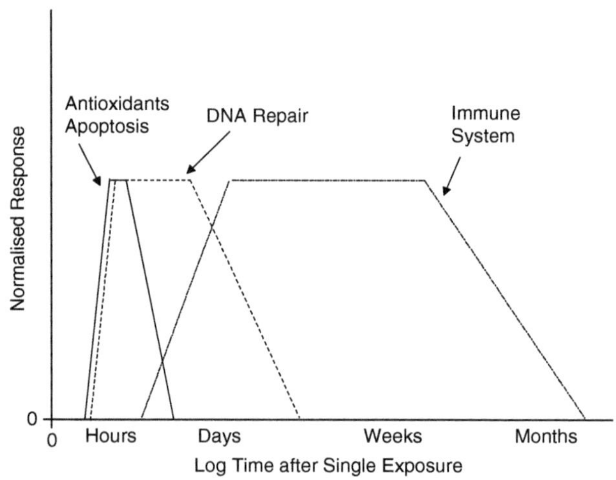


**Fig. 2.3** Temporal stimulation of antioxidants, DNA repair, apoptosis, and the immune system following exposure to ionizing radiation [49]

Adaptation has been observed with both low-LET and high-LET radiations [40]. A combination of a low LET (X-ray) priming dose and a high LET ($\alpha$-particles) challenging dose leads to adaptation in human lymphocytes [50]. Adaptive responses to 1–100 mGy increase cellular DNA double-strand break repair capacity, reduce cell death, reduce chemical- and radiation-induced chromosome aberrations and mutations, and reduce spontaneous and radiation-induced neoplastic transformation in vitro [51].

The resistance to high-dose ionizing radiation acquired by prior low-dose radiation exposure was first identified by Olivieri in human lymphocytes *in vitro* [52]. The optimum low-LET dose for low-dose radiation used to demonstrate the radioadaptive response is <100 mGy

**Table 2.2** Radioadaptive response in human lymphocytes exposed to X-rays

| Dose treatment (mGy) | % Expected chromatid + Isochromatid breaks |
|---|---|
| 10 + 1,500 | 69.8 |
| 50 + 1,500 | 65.9 |
| 100 + 1,500 | 69.5 |
| 200 + 1,500 | 77.6 |
| 300 + 1,500 | 84.6 |
| 400 + 1,500 | 83.7 |
| 500 + 1,500 | 96.2 |

The 10–500 mGy pretreatment doses were given at 32 h after PHA stimulation and the 1,500 mGy dose was given at 48 h after PHA stimulation [55].

*in vitro* [53]. A conditioning dose of 50–100 mGy followed by a challenge dose of 8 Gy (4-h interval) significantly reduced the induction of DNA strand breaks [54]. A conditioning dose of 10–500 mGy reduced the formation of chromosome breaks in cultured human lymphocytes with 10–100 mGy, showing the greatest radioadaptive response (Table 2.2) [55]. A whole body priming dose in vivo in mice of 150–600 mGy protected against birth effects [56].

The radioadaptive response is divided into three successive biological processes: (1) the intracellular response, (2) the extracellular signal, and (3) maintenance. The extracellular signal occurs by release of diffusible signaling molecules and/or by gap-junction intercellular communication. Bystander effects following multiple low-dose exposures implicate adaptive responses induced in cells that receive signals from irradiated neighboring cells. Biological effects arise in those cells that receive no direct radiation exposure [57]. Nontargeted (bystander) effects of ionizing radiation involve mutagenesis, genomic instability, cell differentiation, micronucleus formation, cell transformation, and apoptosis. These may be either detrimental or apoptosis-mediated protector effects [58]. The selective removal of nonhit genomically damaged cells by bystander-induced apoptosis or terminal differentiation is responsible for protective effects [59–61]. Alpha particles result in a bystander effect, which is subject to modulation by the radioadaptive response [62, 63]. Thus, the bystander effect may be a protective response that informs neighboring cells to activate their defenses [64].

Low-dose radiation stimulates antioxidant defense mechanisms, inactivating reactive oxygen species produced by high doses of radiation, and enhancing cellular antioxidant defenses [65, 66]. Cellular metabolism generates the toxic superoxide radical. A family of superoxide dismutases protects against the superoxide radical [67]. A cumulative $\gamma$-ray dose of 0.5 Gy, given at a rate of 1.2 mGy $h^{-1}$, significantly increased the expression of catalase and Mn superoxide dismutase (Fig. 2.4) [68]. The beneficial effect of regular exercise is also due to upregulation of antioxidant defenses and DNA repair [69].

Activated high-fidelity DNA repair by low doses of low-LET radiation contributes to the suppression of stochastic effects. Stimulation of DNA repair by low doses of ionizing radiation is responsible in part for the radioadaptive response [70]. For example, preirradiation of mouse spleen cells with 0.5 Gy given at 1.2 mGy $h^{-1}$ significantly protected against DNA damage from a challenge dose (Fig. 2.5) [68].

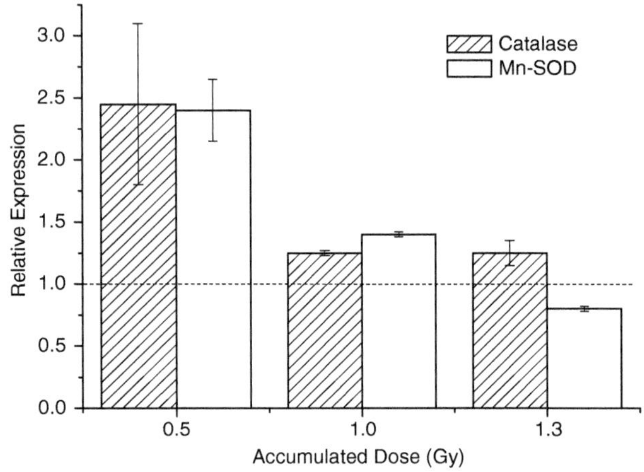


**Fig. 2.4** Antioxidant enzyme expression in mouse spleen cells following chronic γ-irradiation at 1.2 mGy $h^{-1}$ [68]

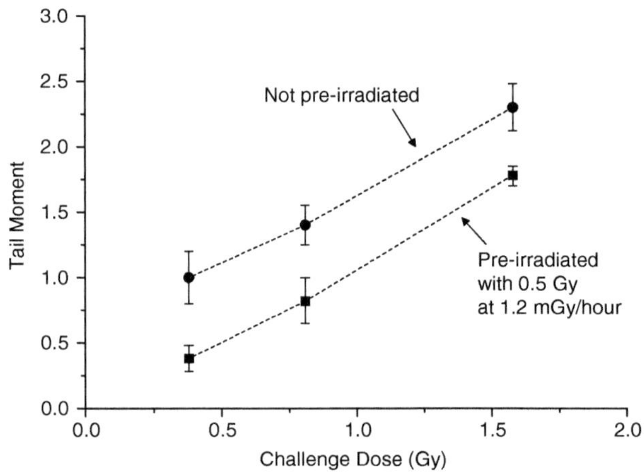


**Fig. 2.5** Results of the comet assay in mouse spleen cells pretreated with 0.5 Gy given at 1.2 mg $h^{-1}$ followed by a challenge dose [68]

## 2.3 Chromosome Aberrations

The radioadaptive response has been observed in studies with human peripheral blood lymphocytes following high-background environmental [71, 72], A-bomb [73], Chernobyl accident [74, 75], and occupational exposures in medical [76, 77] and nuclear workers [78]. Both in vivo and in vitro studies have failed to demonstrate an increase in chromosome aberrations at doses less than 20 mSv [79]. Low doses of radiation enhance the resistance to effects of subsequent higher doses [80]. X-ray doses as low as 5 mGy demonstrated an adaptive response for chromosome aberrations in human lymphocytes [55, 81] or exhibited hormesis by itself (Fig. 2.6) [82].

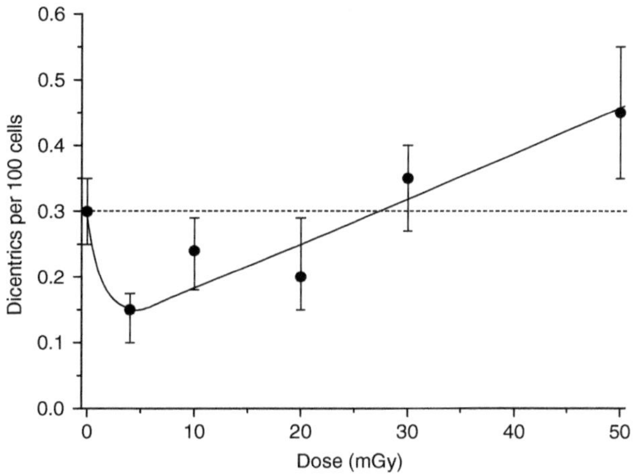


**Fig. 2.6** Dicentric chromosome aberration yield as a function of radiation dose [82]

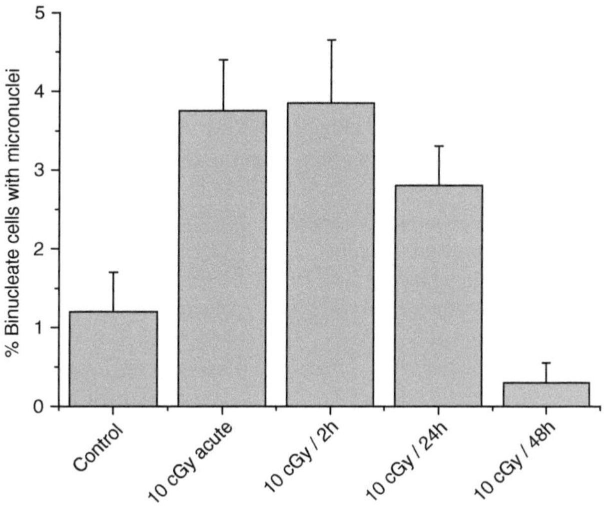


**Fig. 2.7** Influence of dose rate on the frequency of micronucleus formation in human fibroblasts assayed immediately after receiving a cumulative dose of 100 mGy photons from $^{137}$Cs [86]

Chromosome translocations in lymphocytes of former plutonium workers showed a decreasing frequency with increasing cumulative lung doses of up to 1 Sv [83]. An adaptive dose of 0.001 mGy, which was as effective as doses of 0.01, 1, or 10 mGy, followed 4 h later by a challenge dose of 1,000 mGy X-rays, protected against chromosomal inversions in pKZ1 mouse prostate. The 0.001 mGy priming dose was three orders of magnitude lower than had been previously reported [84]. Damage and repair of DNA damage is not consistent among all chromosomes. For example, preferentially repaired breaks were observed in chromosomes 2 and 18 of human fibroblasts, while breaks in chromosome 19 were often left unrepaired; repair of breaks in chromosomes 4 and 7 were unchanged [85]. Protracting exposure of normal human fibroblasts to 10 cGy $^{137}$Cs $\gamma$ rays over a 48-h period reduced micronucleus frequency to a level lower than spontaneous (Fig. 2.7) [86].

## 2.4 Neoplastic Transformation

Low doses of low-LET radiation given at low-dose rates suppressed neoplastic transformation from a high challenge dose (Table 2.3) as well as decreasing transformation rates to below normal spontaneous levels. Doses less than 100 mSv delivered at dose rates of 1–4 mSv day$^{-1}$ induced an adaptive response against neoplastic transformation in vitro. A dose-rate-dependent threshold of <1 mSv day$^{-1}$ for neoplastic suppression was observed [88, 89]. Chronic exposure to γ-ray doses of 1–100 mGy reduced the frequency of neoplastic transformation to levels below the spontaneous rate [90]. High-LET alpha radiation does not activate the radioadaptive response for neoplastic transformation. However, a threshold of above 200 mGy was seen for neoplastic transformation of HeLa x skin fibroblast human hybrid cells exposed to 1 Gev/nucleon iron ions, possibly as a result of the radioadaptive response from associated low-LET δ rays [91].

The resulting dose–response curves are U- or J-shaped with a threshold of less than 100 mSv (Fig. 2.8) [90, 92] and a maximum hormesis response at 1 mGy (Fig. 2.9) [93]. Suppression of neoplastic transformation was similar at 1, 10, and 100 mSv given at 2.4 mSv min$^{-1}$ (Table 2.4) [90]. Thirty kvp photons from $^{125}$I resulted in a threshold of 100 mSv for neoplastic transformation in a human skin fibroblast cell line when given at 0.91 and 1.9 mGy min$^{-1}$ (Fig. 2.10) [88]. A very low-dose rate resulted in significantly less neoplastic transformation than found in unirradiated controls (Fig. 2.11) [89].

**Table 2.3** Radiation-induced adaptive response for protection against neoplastic transformation in C3H 10t1/2 mouse embryo cells [87]

| Treatment | Transformation frequency (× 10$^{-4}$) |
|---|---|
| Unexposed control | 3.7 |
| 4 Gy (high-dose-rate) | 41 |
| 100 mGy (low-dose-rate) + 4 Gy (high-dose-rate) | 16 |

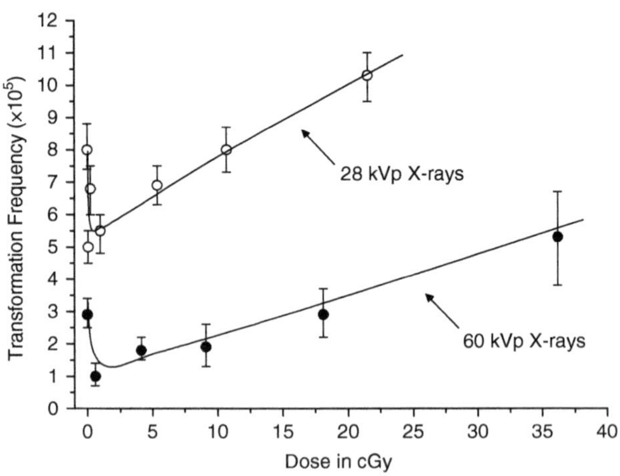


**Fig. 2.8** Influence of dose on neoplastic transformation of human hybrid cells [92]

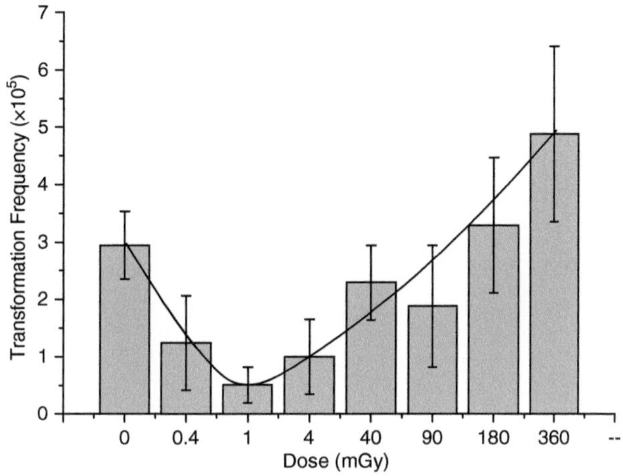


**Fig. 2.9** Neoplastic transformation of HeLa x skin fibroblast hybrid cells treated with 60 kvp X-rays [93]

**Table 2.4** Decrease in neoplastic transformation below the spontaneous rate in C3H 10t1/2 cells by low-dose radiation [90]

| Exposed dose (mGy) at 2.4 mGy min$^{-1}$ | Transformation frequency ($\times 10^{-3}$) | Percent of control |
|---|---|---|
| Unexposed control | 1.8 | – |
| 1.0 | 0.62 | 34 |
| 10 | 0.39 | 22 |
| 100 | 0.49 | 27 |

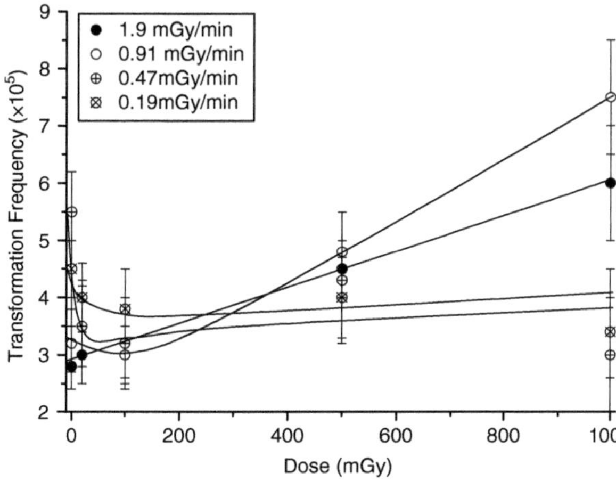


**Fig. 2.10** Influence of dose rate (30 keV $^{125}$I) on neoplastic transformation [88]

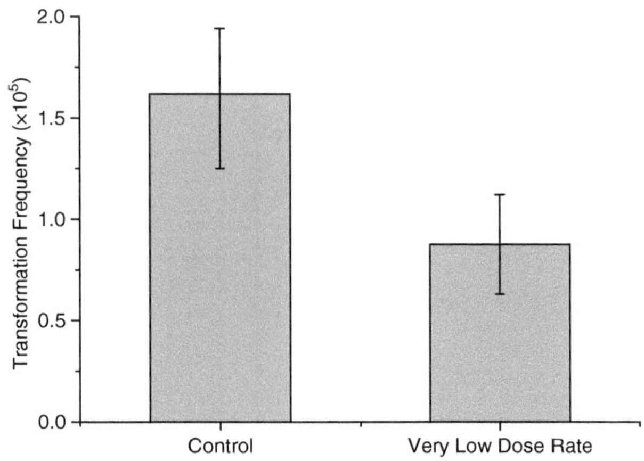


**Fig. 2.11** Neoplastic transformation of HeLa x skin fibroblast hybrid cells at a very low dose rate (VLDR) of ~2 mGy day$^{-1}$ (30 kvp photons) for an accumulated dose of 194 mGy [89]

Genomically damaged (chromosomal aberrations or aneuploidy) and/or neoplastically transformed cells may be selectively killed by apoptosis at low doses of radiation. However, doses less than 250 mSv of low LET radiation do not cause genetic instability [94, 95]. Based on the demonstrated close similarity between the relative risk (RR) dose–response relationships for radiation-induced neoplastic transformation in vitro and cancer induction in humans, a novel, RR model for neoplastic transformation was adapted for application to RR estimation in irradiated human populations [96]. The model involves different thresholds for activating and inhibiting the hormetic response in humans. Low doses activate (stimulate) whereas moderate and high doses inhibit [97].

## 2.5 Apoptosis

Cells carry in their genome a program for self-destruction called apoptosis. Apoptosis, or programmed cell death, limits the accumulation of potentially harmful cells. Failure to regulate tissue homeostasis by apoptosis results in cancer development [98]. Apoptosis is morphologically characterized by nuclear condensation and fragmentation (pyknosis), plasma membrane blebbing, and cell shrinkage. Defective apoptosis allows tumor cells to better survive hypoxia. In addition, failed apoptosis allows more time for transformed genetically unstable cells to become more malignant, invasive, and metastatic. Defective apoptosis plays a role in resistance to chemotherapy and radiotherapy for cancer, increasing the threshold for cell killing effectiveness and therapeutic doses for tumor-cell killing [99, 100].

Error-prone repair can lead to cancer formation (Fig. 2.12). Cell regulators, such as growth factors, cytokines and hormones, are involved with the activation and repression of genes associated with apoptosis. Apoptosis is enhanced by low to moderate doses of ionizing radiation in normal tissues and tumors [102]. Apoptosis in transformation sensitive G2/M phase subpopulation of cells could account for reduced cell transformation and cancer incidence for low doses of radiation [103].

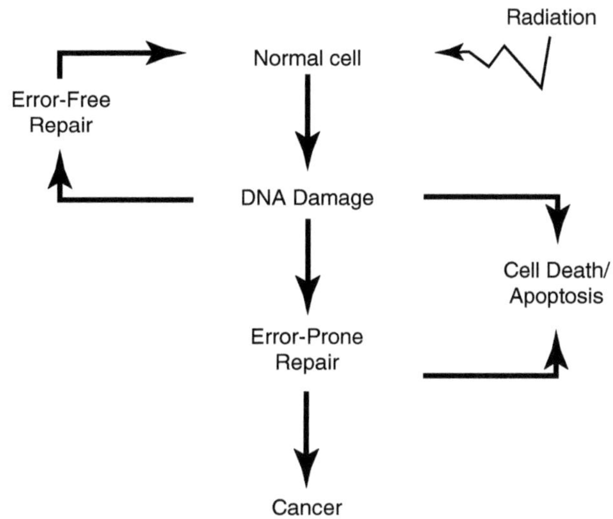


**Fig. 2.12** Possible roles of error-free and error-prone repair and apoptosis in cancer formation [101]

Low dose and dose rate of low-LET radiation above a stochastic threshold dose stimulates intra- and intercellular signaling, leading to Activated Natural Protection (ANP) against genomic-instability-associated disease, such as cancer [46, 104]. Mechanistic pathways for low-dose radiation ANP may include induced p53-dependent high-fidelity DNA repair along with normal apoptosis activation of an epigenetic protective apoptosis-mediated (PAM) process that selectively removes premalignant transformed cells at doses as low as about 1 mSv [61, 105], while also enhancing immune functions [106, 107].

The PAM process along with stimulation of the immune system limits potential cancer formation, when activated by $\gamma$- and X-radiations [97]. Stimulation of the PAM process by low-dose, low-LET radiation may cause apoptosis of cells transformed by the other agents, including chemical carcinogens in cigarette smoke, decreasing lung cancer risk [108]. A novel biological-based hormetic relative risk (HRR) model for cancer induction by low-dose, low-dose-rate irradiation has been proposed out of concern about the inappropriateness of the LNT assumption [109].

Mice with a knockout *Fas* gene that controls apoptosis have a maximum lifespan of only 150 days. Lifespan chronic $\gamma$-irradiation at 0.35 mGy $h^{-1}$ increased maximum lifespan to 280 days. Increasing the lifespan dose-rate to 1.2 mGy nearly replaced lost apoptosis function of the *Fas* gene, maintaining a near normal lifespan (Fig. 2.13) [110]. This study clearly demonstrated the importance of apoptosis in minimizing cancer risk by chronic irradiation.

Apoptosis of chemically transformed cells may be activated by low-dose, low-LET $\gamma$- or X-radiations [111–113]. These studies indicate that low LET radiation increases apoptosis, removing cigarette-smoke and high-LET radiation induced, genomically damaged pulmonary cells before they can develop into lung cancer [114, 115]. A dose as low as 2 mGy $\gamma$-rays or 0.3 mGy $\alpha$-particles is sufficient to produce an observable response in apoptosis in unirradiated transformed cells cocultured with irradiated cells [116].

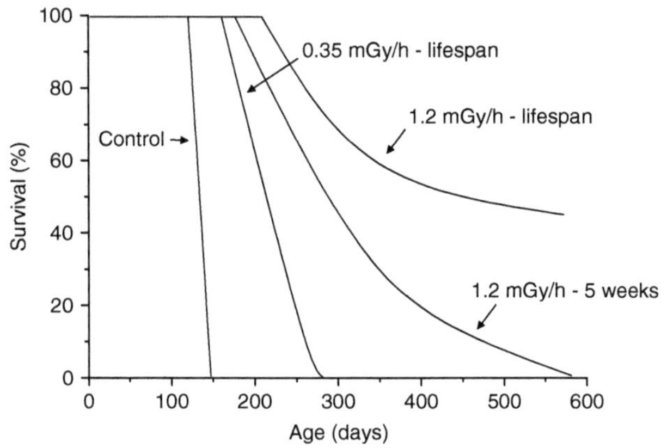


**Fig. 2.13** Lifespan in mice carrying a deletion in the apoptosis-regulating *Fas* gene following chronic γ-irradiation [110]

## 2.6 Immune Enhancement

Two categories of immune cells cooperate to defend against pathogenic microorganisms and cancer. They comprise the innate and adaptive immune responses. Low-dose ionizing radiation stimulates the immune system in vivo and in vitro (Fig. 2.14) [117–128], in part, by mobilizing hematopoietic progenitor cells into the peripheral blood [129].

Low-dose radiation enhances the adaptive immune response (Table 2.5) [106, 107, 118–127, 130]. The secretion of the proinflammatory cytokines IL-12 and IL-18 by mouse macrophages is increased by a dose of only 75 mGy [131]. These cytokines participate in the adaptive immune response. Full activation of the adaptive immune response to malignant cells might facilitate their potential eradication.

Cytokine production is increased by low doses of whole-body irradiation, followed by a decrease at higher doses (Fig. 2.15) [132]. Single, whole-body, X-ray exposures of mice to 100 or 200 mGy significantly stimulated the production of the cytokines IL-1β, IL-2, IL-12, TNF-α, and IFN-γ by macrophages and splenocytes. A dose of 200 mGy was a

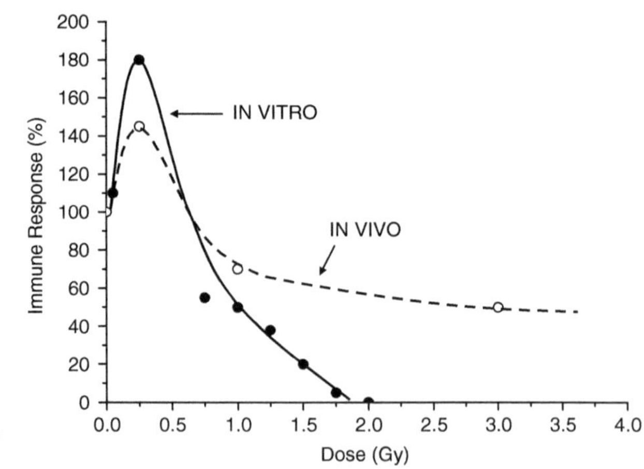


**Fig. 2.14** Immune response to anti-sheep RBC by splenic cells in mice following exposure to ionizing radiation [117]

**Table 2.5** Low-dose X-irradiation (75 mSv) stimulates cellular immunity in mice [107]

| Parameter | Change (%) | *p* value |
|---|---|---|
| NK activity | +19 | <0.05 |
| Macrophage activity | +52 | <0.05 |
| Cytotoxic T Lymphocytes | +40 | <0.01 |
| Antibody-dependent cell-mediated cytotoxicity | +30 | <0.05 |
| T-cell proliferation | +101 | <0.01 |

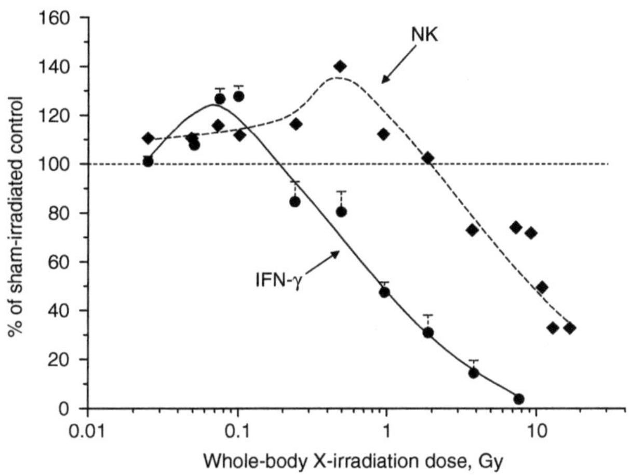


**Fig. 2.15** NK activity and IFN-γ secretion by mouse splenocytes following whole-body x-irradiation [132]

greater stimulant of cytokine production than was 100 mGy [133]. These cytokines are known for their direct and indirect antitumor effects.

Low-dose radiotherapy has been used to treat a variety of inflammatory conditions [134, 135]. A significant antiinflammatory effect was seen in mice following a dose of 300 mGy from 6 MeV photons, which maximally inhibited leukocyte recruitment [136]. A dose of 0.5 Gy γ radiation causes an upregulation of MKP-1, leading to inactivation of p38 MAPK and suppression of TNF-α production in mouse macrophages, which inhibited inflammation [137]. Mice preexposed to 75 mSv X-rays, 6 h before tumor implantation, exhibited larger areas of tumor necrosis and infiltration with lymphocytes than in the nonradiation group. Endogenously administered $H_2O_2$ is beneficial in immune stimulation, acting as a second messenger in lymphocyte activation [138]. Erythrocyte immune function and SOD activity were also significantly increased in the low-dose radiation mice [139].

Whole-body exposure to low doses of ionizing radiation stimulated immune function in humans [30, 31, 140–142]. Villagers from Ramsar, Iran, receiving a background dose of 260 mSv year$^{-1}$, exhibited an increased lymphocyte-induced IL-4 and IL-10 production. IL-2 and IL-4 production by peripheral blood lymphocytes was also increased in X-ray equipment operators. Lymphocyte proliferation in response to PHA was unaffected [143].

## References


1. Mayneord W (1964) Radiation and health. The Nuffield Provincial Hospital Trust. London, UK
2. Pollycove M (2002) Radiation hormesis: the biological response to low doses of ionizing radiation. Health Effects of Low-Level Radiation, BNES
3. Billen D (1990) Spontaneous DNA damage and its significance for the "negligible dose" controversy in radiation protection. Radiat Res 124:242–245
4. Stewart RD (1999) On the complexity of the DNA damages created by endogenous processes. Radiat Res 152:101–105
5. Mosmann KL (2007) Radiation risk in perspective. CRC, Taylor and Francis, Boca Raton, FL
6. Satta L et al (2002) Influence of a low background radiation environment on biochemical and biological responses in V79 cells. Radiat Environ Biophys 41:217–224
7. Muller HJ (1927) Artificial transmutation of the gene. Science 116:84–87
8. Oliver CP (1930) The effect of varying the duration of X-ray treatment upon the frequency of mutation. Science 121:44–46
9. ICRP, Recommendations of the International Commission on Radiological Protection, as amended 1959 and revised 1962 (1964) Publication 6, International Commission on Radiological Protection, Pergamon Press, Oxford
10. UNSCEAR (1993) United Nations Scientific Committee on the Effects of Atomic Radiation. Sources and Effects of Ionizing Radiation. UNSCEAR 1993 Report to the General Assembly with Scientific Annexes, United Nations, NY
11. UNSCEAR (2001) Hereditary effects of radiation. Scientific annex of UNSCEAR 2001 report to the General Assembly, United Nations Scientific Committee on the Effects of Atomic Radiation, Vienna, Austria, p 224
12. Koana T, Takashima Y, Okada MO et al (2004) A threshold exists in the dose-response relationship for somatic mutation frequency induced by X irradiation of Drosophila. Radiat Res 161:391–396
13. Koana T, Takashima Y, Okada O et al (2007) Reduction of background mutations by low-dose x irradiation of Drosophila spermatocytes at a low dose rate. Radiat Res 157:217–221
14. Ogura K, Magae J, Kawakami Y, Koana T (2009) Reduction in mutation frequency by very low-dose gamma irradiation of Drosophila melanogaster germ lines. Radiat Res 171:1–8
15. Koana T, Okada MO, Ogura K (2007) Reduction of the background mutation by a low dose irradiation of Drosophila spermatocytes at a low dose-rate. 6th LOWRAD Conference, Budapest, Hungary. Abstract, p 72
16. Yang F, Stenoien DI, Strittmatter EF et al (2006) Phosphoproteome profiling of human skin fibroblast cells in response to low- and high-dose radiation. J Proteome Res 5:1252–1260
17. Amundson A, Lee RA, Koch-Paiz CA et al (2003) Differential responses of stress genes to low dose-rate gamma irradiation. Mol Cancer Res 1:445–452
18. Franco N, Lamatine J, Frouin V et al (2005) Low-dose exposure to g rays induces specific gene regulations in normal human keratinocytes. Radiat Res 163:623–635
19. Magae J, Hoshi Y, Furukawa C et al (2003) Quantitative analysis of biological responses to ionizing radiation, including dose, irradiation time, and dose rate. Radiat Res 160:543–548
20. Nakamura H, Fukami H, Hayashi Y et al (2005) Cytotoxic and mutagenic effects of chronic low-dose-rate irradiation on TERT-immortalized human cells. Radiat Res 163:283–288
21. Tubiana M, Arengo A, Averbeck D, Masse R (2007) Low-dose risk assessment: comments on the summary of the international workshop. Radiat Res 167:742–744
22. Saul RL, Ames BN (1986) Background levels of DNA damage in the population. Basic Life Sci 38:529–535

23. Little JB (2000) Radiation carcinogenesis. Carcinogenesis 21:397–404
24. Tubiana M, Aurengo A, Averbeck D, Masse R (2008) Letters to the Editor. Low-dose risk assessment: the debate continues. Radiat Res 169:245–248
25. Leonard BE (2005) Adaptive response by single cell radiation hits-Implications for nuclear workers. Radiat Prot Dosimetry 116:387–391
26. Leonard BE (2007) Adaptive response and human benefit: Part I – a microdosimetry dose dependent model. Int J Radiat Biol 83:115–131
27. Leonard BE (2007) Adaptive response: Part II – Modeling for dose rate and time influences. Int J Radiat Biol 83:395–408
28. Roots R, Chatterjee A, Chang P et al (1985) Characterization of hydroxyl radical-induced damage after sparsely and densely ionizing radiation. Int J Radiat Biol 47:157–166
29. Calabrese EJ, Baldwin L (2001) Scientific foundations of hormesis. Crit Rev Toxicol 31:351–624
30. Luckey TD (1980) Hormesis with ionizing radiation. CRC Press, Boca Raton, FL
31. Luckey TD (1982) Physiological benefits from low-level ionizing radiation. Health Phys 43:771–789
32. Wallace SS (1988) AP-endonucleases and DNA-glycosylases that recognize oxidative DNA damage. Environ Mol Mutagen 12:431–477
33. Luckey TD (1991) Radiation hormesis. CRC Press, Boca Raton, FL
34. Luckey TD (1999) Nurture with ionizing radiation: a provocative assumption. Nutr Cancer 34:1–11
35. Pollycove M, Feinendegen LE (2001) Biologic responses to low doses of ionizing radiation: Deteriment versus hormesis. Part 2: Dose responses to organisms. J Nucl Med 42: 26N–37N
36. Pollycove M, Feinendegen LE (1999) Molecular biology, epidemiology and the demise of the linear no-threshold (LNT) assumption. C R Acad Sci Paris Life Sci 322:197–204
37. Pollycove M (1998) Nonlinearity of radiation health effects. Environ Health Perspect 106:363–368
38. Gonzalgo ML, Jones PA (1997) Mutagenic and epigenetic effects of DNA methylation. Mutat Res 386:107–118
39. Esteller M (2005) Aberrant DNA methylation as a cancer-inducing mechanism. Ann Rev Pharmacol Toxicol 45:629–656
40. Tapio S, Jacob V (2007) Radioadaptive response revisited. Radiat Enviorn Biophys 46:1–12
41. Aurengo A, Averbeck D, Bonnin A et al (2005) Dose-effect relationships and estimation of the carcinogenic effects of low doses of ionizing radiation. Executive Summary. French Academy of Sciences, French National Academy of Medicine
42. Bhattacherjee G, Ito A (2001) Deceleration of carcinogenic potential by adaptation with low dose gamma irradiation. In Vivo 15:87–92
43. Ibuki Y, Goto R (1994) Adaptive response to low doses of gamma-ray in Chinese hamster cells: determined by cell survival and DNA synthesis. Biol Pharm Bull 17:1111–1113
44. Ikushima T, Aritomi H, Morisita J (1996) Radioadaptive response: Efficient repair of radiation-induced DNA damage in adapted cells. Mutat Res 358:193–198
45. Scott BR (2007) Low-dose radiation-induced protective process and implications for risk assessment, cancer prevention, and cancer therapy. Dose Response 5:131–141
46. Scott BR (2005) Stochastic thresholds: a novel explanation of nonlinear dose-response relationships. Dose Response 3:547–567
47. Collis SJ, Schwaninger JM, Ntambi AJ et al (2004) Evasion of early cellular response mechanisms following low level radiation-induced DNA damage. J Biol Chem 279: 49624–49632
48. Kant K, Chauhan RP, Sharma GS et al (2003) Hormesis in humans exposed to low-level ionizing radiation. Int J Low Radiat 1:76–87

49. Feinendegen LE, Paratzke HG, Neumann RD (2007) Damage propagation in complex biological systems following exposure to low doses of ionising radiation. Atoms for Peace: An International Journal 1:336–354
50. Wolff S, Afzal V, Jostes RF et al (1993) Indications of repair of radon-induced chromosome damage in human lymphocytes: an adaptive response induced by low doses of X-rays. Environ Health Perspect 101(Suppl 3):73–77
51. Mitchel REJ (2006) Cancer and low dose response in vivo: implications for radiation protection. Can Nucl Soc Bull 27:23–26
52. Olivieri G, Bodycote J, Wolff S (1984) Adaptive response of human lymphocytes to low concentrations of radioactive thymidine. Science 223:594–597
53. Sasaki MS (1995) On the reaction kinetics of the radioadaptive response in cultured mouse cells. Int J Radiat Biol 68:281–291
54. Cramers P, Atanasova P, Vrolijk H et al (2005) Pre-exposure to low doses: modulation of X-ray-induced DNA damage and repair? Radiat Res 164:383–390
55. Shadley JD, Wolff S (1987) Very low doses of X-rays can cause human lymphocytes to become less susceptible to ionizing radiation. Mutagenesis 2:95–96
56. Wang B, Ohyama H, Shang Y et al (2004) Adaptive response in embryogenesis: V. Existence of two efficient dose-rate ranges for 0.3 Gy of priming irradiation to adapt mouse fetuses. Radiat Res 161:264–272
57. Little JB (2006) Lauriston S. Taylor lecture: nontargeted effects of radiation: implications for low-dose exposures. Health Phys 91:416–426
58. Schollnberger H, Mitchel REJ, Redpath JL et al (2007) Detrimental and protective bystander effects: A model approach. Radiat Res 168:614–626
59. Belyakov OV, Folkard M, Mothersill C et al (2002) Bystander-induced apoptosis and premature differentiation in primary urothelial explants after charged particle microbeam irradiation. Radiat Prot Dosimetry 99:249–251
60. Lyng FM, Seymour CB, Mothersill C (2000) Production of a signal by irradiated cells which leads to a response in unirradiated cells characteristic of initiation of apoptosis. Br J Cancer 83:1223–1230
61. Portess DI, Bauer G, Hill MA et al (2007) Low dose irradiation of non-transformed cells stimulates the selective removal of pre-cancerous cells via intercellular induction of apoptosis. Cancer Research 67:1246–1253
62. Zhou H, Randers-Pehrson G, Geard CR et al (2003) Interaction between radiation-induced adaptive response and bystander mutagenesis in mammalian cells. Radiat Res 160:512–516
63. Leonard BE (2008) Letters to the Editor. Common sense about the linear no-threshold controversy-give the general public a break. Radiat Res 169:245–248
64. Mothersill C, Seymour CB (2006) Radiation-induced bystander effects and the DNA paradigm: an "out of field" perspective. Mutat Res 59:5–10
65. Laval F (1988) Pretreatment with oxygen species increases the resistance of mammalian cells to hydrogen peroxide and gamma rays. Mutat Res 201:73–79
66. Kojima S, Ishida H, Takahashi M et al (2002) Elevation of glutathione induced by low-dose gamma rays and its involvement in increased natural killer activity. Radiat Res 157:275–280
67. McCord JM (2008) Superoxide dismutase, lipid peroxidation, and bell-shaped dose response curves. Dose Response 6:223–238
68. Otsuka k, Koana T, Tauchi H, Sakai K (2006) Activation of antioxidative enzymes induced by low-dose-rate whole-body $\gamma$ irradiation: adaptive response in terms of initial DNA damage. Radiat Res 166:474–478
69. Radak Z, Chung HY, Goto S (2005) Exercise and hormesis: oxidative stress-related adaptation for successful aging. Biogerontology 6:71–75
70. Pant MC, X-Y Liao QLu et al (2003) Mechanisms of suppression of neoplastic transformation in vitro by low doses of low LET radiation. Carcinogenesis 24:1961–1965

71. Ghiassi-nejad M, Mortazavi SM, Cameron JR et al (2002) Very high background radiation areas of Ramsar, Iran: preliminary biological studies. Health Phys 82:87–93
72. Masoomi JR, Mohammadi S, Amini M et al (2006) High background radiation areas of Ramsar in Iran: evaluation of DNA damage by alkaline single cell gel electrophoresis (SCGE). J Environ Radioact 86:176–186
73. Gajendiran N, Tanaka K, Kumaravel TS et al (2001) Neutron-induced adaptive response studied in G0 human lymphocytes using the comet assay. J Radiat Res 42:91–101
74. Padovani L, Appolloni M, Anzidei P et al (1995) Do human lymphocytes exposed to the fallout of Chernobyl accident exhibit an adaptive response? 1. Challenge with ionizing radiation. Mutat Res 332:33–38
75. Tedeschi B, Caporossi D, Vernole P et al (1995) Do human lymphocytes exposed to the fallout of Chernobyl accident exhibit an adaptive response? 2. Challenge with bleomycin. Mutat Res 332:39–44
76. Barquinero JF, Barrios L, Caballin MR et al (1995) Occupational exposure to radiation induces an adaptive response in human lymphocytes. Int J Radiat Biol 67:187–191
77. Gourabi H, Mozdarani H (1998) A cytokinesis-blocked micronucleus study of the radioadaptive response of lymphocytes of individuals occupationally exposed to chronic doses of radiation. Mutagenesis 13:475–480
78. Thierens H, Vral A, Barbe M et al (2002) Chromosomal radiosensitivity study of temporary nuclear workers and the support of the adaptive response induced by occupational exposure. Int J Radiat Biol 78:1117–1126
79. United Nations (2000) United Nations Scientific Committee on the effects of atomic radiation. Sources, effects and risks of ionizing radiation. United Nations, New York
80. Grillo CA, Dulout FN, Guerci AM (2009) Evaluation of radioadaptive response induced in CH-K1 cells in a non-traditional model. Int J Radiat Biol 85:159–166
81. Shadley JD (1994) Chromosomal adaptive response in human lymphocytes. Radiat Res 138(Suppl):S9–S12
82. Pohl-Ruling J, Fischer P, Haas O et al (1983) Effect of low-dose acute x-irradiation on the frequencies of chromosomal aberrations in human peripheral lymphocytes in vitro. Mut Res 110:71–82
83. Livingston GK, Falk RB, Schmid E (2006) Effect of occupational radiation exposures on chromosome aberration rates in former plutonium workers. Radiat Res 166:89–97
84. Day TK, Zeng G, Hooker AM et al (2006) Extremely low priming doses of X radiation induce an adaptive response for chromosomal inversions in pKZ1 mouse prostate. Radiat Res 166:757–766
85. Broome EJ, Brown DL, Mitchell REJ (1999) Adaptation to radiation alters biases in DNA repair at the chromosome level. Int J Radiat Biol 75:681–690
86. de Toledo SM, Asaad N, Ventkatachalam P et al (2006) Adaptive responses to low-dose/low-dose-rate g rays in normal human fibroblasts: the role of growth architecture and oxidative metabolism. Radiat Res 166:849–857
87. Azzam EI, Raaphorst GP, Mitchel REJ (1994) Radiation-induced adaptive response for protection against micronucleus formation and neoplastic transformation in C3H 10t1/2 mouse embryo cells. Radiat Res 138:S28–S31
88. Elmore E, Lao X-Y, Kapadia R et al (2006) The effect of dose rate on radiation-induced neoplastic transformation in vitro by low doses of low-LET radiation. Radiat Res 166: 832–838
89. Elmore E, Lao X-Y, Kapadia R et al (2008) Low doses of very low-dose-rate low-LET radiation suppress radiation-induced neoplastic transformation in vitro and induce an adaptive response. Radiat Res 169:311–318
90. Azzam EI, De Toledo SM, Raaphorst GP et al (1996) Low-dose ionizing radiation decreases the frequency of neoplastic transformation to a level below the spontaneous rate in C3H 10T1/2 cells. Radiat Res 146:369–373

91. Elmore E, Lao X-Y, Kapadia R, Redpath JL (2009) Threshold-type dose response for induction of neoplastic transformation by 1 GeV/nucelon iron ions. Radiat Res 171:764–770
92. Redpath JL, Lu Q, Lao X-Y et al (2003) Low doses of diagnostic energy X-rays protect against neoplastic transformation in vitro. Int J Radiat Biol 79:235–240
93. Redpath JL (2006) Suppression of neoplastic transformation in vitro by low doses of low let radiation. Dose Response 4:302–308
94. Kadhim MA, Hill MA, Moore SR (2006) Genomic instability and the role of radiation quality. Radiat Prot Dosimetry 122:221–227
95. Okada M, Okabe A, Uchihori Y et al (2007) Single extreme low dose/low dose rate irradiation causes alteration in lifespan and genome instability in primary human cells. Br J Cancer 96:1707–1710
96. Redpath JL, Liang D, Taylor TH et al (2001) The shape of the of the dose-response curve for radiation-induced neoplastic transformation. Evidence for an adaptive response against neoplastic transformation at low doses of low-LET radiation. Radiat Res 156:700–707
97. Scott BR (2007) Low-dose radiation-induced protective process and implications for risk assessment, cancer prevention, and cancer therapy. Dose Response 5:131–141
98. Thompson HJ, Strange R, Schedin PJ (1992) Apoptosis in the genesis and prevention of cancer. Cancer Epidemiol Biomarkers prev 1:597–602
99. Reed JC (2003) Apoptosis-targeted therapies for cancer. Cancer Cell 3:17–22
100. Reed JC, Doctor KS, Godzik A The domains of apoptosis: A genomics perspective. Science STKE 29 June 2004/239,/re9.
101. Mitchel REJ (2007) Low doses of radiation reduce risk in vivo. Dose Response 5:1–10
102. Stephens LC, Kang K, Schultheiss TE et al (1991) Apoptosis in irradiated murine tumors. Radiat Res 127:308–316
103. Redpath JL, Short SC, Woodcock M et al (2003) Low-dose reduction in transformation frequency compared to unirradiated controls: the role of hyper-radiosensitivity to cell death. Radiat Res 159:433–436
104. Scott BR, JDi Palma (2006) Sparsely ionizing diagnostic and natural background radiations are likely preventing cancer and other genomic-instability-associated disease. Dose Response 5:230–255
105. Bauer G (2007) Low dose radiation and intercellular induction of apoptosis: potential implications for control of oncogenesis. Int J Radiat Biol. 83:873–888
106. Liu SZ (2003) Nonlinear dose-response relationship in the immune system following exposure to ionizing radiation: mechanisms and implications. Nonlinearity Biol Toxicol Med 1:71–92
107. Liu S (2007) Cancer control related to stimulation of immunity by low-dose radiation. Dose Response 5:39–47
108. Sanders CL, Scott BR (2008) Smoking and hormesis as confounding factors in radiation pulmonary carcinogenesis. Dose Response 6:53–79
109. Scott BR (2007) Hormesis and the control of genomic instability. In: New Research on Genomic Instability (preliminary title). Nova Science Publishers, Inc. Hauppage, NY (in press).
110. Ina Y, Sakai K (2005) Further study of prolongation of life span associated with immunological modification by chronic low-dose-rate irradiation in MRL-lpr/lpr mice. Effects of whole-life irradiation. Radiat Res 163:418–423
111. Mitchel REJ, Gragtmans NJ, Morrison DP (1999) Beta-radiation-induced resistance to MNNG initiation of papilloma but not carcinoma formation in mouse skin. Radiat Res 121:180–186
112. Sakai K, Hoshi Y, Nomura T et al (2003) Suppression of carcinogenic processes in mice by chronic low dose rat gamma-irradiation. Int J Low Radiat 1:142–146
113. Monchaux G, Morlier JP, Morin M et al (1994). Carcinogenic and cocarcinogenis effects of radon and radon daughters in rats. Environ Health Perspect 102:64–7

114. Stephens LC, Kang K, Schultheiss TE et al (1991) Apoptosis in irradiated murine tumors. Radiat Res 127:308–316;
115. Thompson HJ, Strange R, Schedin PJ (1992) Apoptosis in the genesis and prevention of cancer. Cancer Epidemiol Biomarkers Prev 1:597–602
116. Hill MA (2007) The relevance of radiation track structure at low dose and dose rates. 6th LOWRAD Conference, Budapest, Hungary. Abstract, p 59
117. Mackinodan T, James SJ (1990) T cell potentiation by low dose ionizing radiation: possible mechanisms. Health Phys 59:29–34
118. Taliafero WH, Taliafero LG (1969) Effects of radiation on the initial and anamnestic IgM hemolysin responses in rabbits: antigen injection after X-rays. J Immunol 103:559–569
119. Taliafero WH, Taliafero LG (1970) Effects of irradiation on initial and anamnestic hemolysin responses in rabbits: antigen injection before X-rays. J Immunol 104:1364–1376
120. Hoffsten PE, Dixon FJ (1974) Effect of irradiation and cyclophosphamide on antiKLH antibody formation in mice. J Immunol 112:564–572
121. Anderson RE, Lefkovitz I (1979) In vitro evaluation of radiation-induced augmentation of the immune response. Am J Pathol 97:456–472
122. Anderson RE, Lefkovitz I (1980) Effects of irradiation on the in vitro immune response. Exp Cell Biol 48:255–278
123. Anderson RE, Tokuda S, Williams WL, Spellman CW (1986) Low dose irradiation permits immunization of A/J mice with subimmunogenic numbers of Sal cells. Brit J Cancer 54:505
124. Liu SZ, Liu WH, Sun JB (1987) Radiation hormesis: its expression in the immune system. Health Phys 52:579–583
125. Liu SZ, YC Zhang YMu et al (1996) Thymocyte apoptosis in response to low-dose radiation. Mutation Res 358:185–191
126. Yu Y, Greenstock CL, Trivedi A, Mitchel REJ (1996) Occupational levels of radiation exposure induce surface expression of interleukin-2 receptors in stimulated human peripheral blood lymphocytes. Rad Environ Biophys 35:89–93
127. Ibuki Y, Goto R (1994) Enhancement of cocoavalin A-induced proliferation of splenolymphocytes by low-dose-irradiated macrophages. J Rad Res 35:83–91
128. Upton AC (2001) Radiation hormesis: data and interpretations. Crit Rev Toxicol 31:681–695
129. Li W, Wang G, Chi J et al (2004) Low-dose radiation (LDR) induces hematopoietic hormesis: LDR-induced mobilization of hematopoietic progenitor cells into peripheral blood circulation. Exper Hematol 32:1088–1096
130. Liu XD, Liu SZ, Ma SM, Liu Y (2001) Opposite changes of Il-10 and IL-12 expression in mice after low dose whole-body X-irradiation. J Radiat Res Radiat Prot 19:253–258
131. Shan Y-X, Jin S-Z, Liu X-D et al (2007) Ionizing radiation stimulates secretion of pro-inflammatory cytokines: dose-response relationship, mechanisms and implications. Radiat Environ Biophys 46:21–29
132. Fan XH, Liu SZ JNBUMS 1989, 15:551; YG Yang, SZ LIU. JNBUMS 1989, 15(Suppl):11
133. Cheda A, Nowosielska EM, Wrembel-Wargocka J, Janiak MK (2008) Production of cytokines by peritoneal macrophages and splenocytes after exposures of mice to low doses of X-rays. Radiat Enviorn Biophys 47:275–283
134. Seegenschmiedt MH, Katalinic A, Makoski HB et al (1999) Radiotherapy of benign diseases: A pattern of care study in Germany. Strahlenther Onkol 175:541–547
135. Micke O, Seegenschmiedt MH (2002) Consensus guidelines for radiation therapy of benign diseases: a multicenter approach in Germany. Int J Radiat Oncol Biol Phys 52: 496–513
136. Arenas M, Gil F, Gironella M et al (2006) Anti-inflammatory effects of low-dose radiotherapy in an experimental model of systemic inflammation in mice. Int J Radiat Oncol Biol Phys 66:560–567

137. Tsukimoto M, Homma T, Mutou Y, Kojima S (2009) 0.5 Gy gamma radiation suppresses production of TNF-α through up-regulation of MKP-1 in mouse macrophage RAW264.7 cells. Radiat Res 171:219–224
138. Reth M (2002) Hydrogen peroxide as second messenger in lymphocyte activation. Nature Immunol 3:1129–1134
139. Yu H-S, Song A-Q, Lu Y-D et al (2004) Effects of low-dose radiation on tumor growth, erythrocyte immune function and SOD activity in tumor-bearing mice. Chinese Med J 117:1036–1039
140. Takahashi M, Kojima S, Yamaoka K et al (2000) Prevention of type I diabetes by low-dose gamma irradiation in NOD mice. Radiat Res 154:680–685
141. Luckey TD (1997) Low-dose irradiation reduces cancer death. Rad Prot Manage 14:58–64
142. Hrycek A, Czernecka-Micinski A, Klucinski P et al (2002) Peripheral blood lymphocytes and selected serum interleukins in workers operating X-ray equipment. Toxicol Lett 132: 101–107
143. Attar M, Kondolousy YM, Khansari N (2007) Effect of high dose natural ionizing radiation on the immune system of the exposed residents of Ramsar Town, Iran. Iran J Allergy Asthma Immunol 6:73–78
144. Tubiana M, Feinendegen LE, Yang C, Kaminski JM (2009) The linear no-threshold relationship is inconsistent with radiation biologic and experimental data. Radiology 251:13–22

# Natural Environmental Radiation 3

The disconnect between policy and science causes public confusion and loss of credibility. The LNT makes radiation seem uniquely fearsome and the price we pay is horrendous

(Jim Muckerheide)

Studies of human populations exposed to high environmental levels of low-dose-rate ionizing radiation found in high-background radiation areas (HBRAs) are very useful in predicting disease risks from chronic radiation exposures in nuclear and medical workers and in other exposed populations [1]. Significant doses are seen from naturally occurring radionuclides, particularly from $^{40}$K, and $^{226}$Ra and $^{222}$Rn from uranium decay (Table 3.1). Natural $^{40}$K provides about 380 million disintegrations per day, mostly as β-particles, in standard man. The average whole-body dose rate from $^{40}$K is 0.15–0.20 mSv/year, whereas the annual, average equivalent dose rate from inhaled radon is 1.3 mSv [3, 4]. The vast majority of the earth's surface experiences annual background-radiation levels that range from <1 to >100 mSv. About 99% of the world's population receives dose rates of <7 mSv/year [5]. A few geographical hotspots receive annual doses of >100 mSv/year (Table 3.2) [2]. The global average background radiation dose is 2.4 mSv/year [7]. Residents near Yangjiang in Guangdong province, China, receive an annual background dose of 6.4 mSv [8]. The background-radiation dose in Korea averages 2.5 mSv/year with little variability, whereas the background-radiation dose in the United States averages 2.4 mSv/year but with high variability, ranging from 1.2 to 12 mSv/year.

About 39% of the U.S. population background dose is low LET, whereas 52% of dose is from high-LET radon progeny (mostly $^{222}$Rn) [9]. The annual dose contribution in the United States from nuclear weapon tests (<0.05 mSv) and the Chernobyl accident (<0.02 mSv) are small fractions of the natural, world average, background dose [6].

High maximum annual doses of up to several hundred mSv are found in Kerala, India, and Ramsar, Iran (Table 3.2) [1, 10]. The monazite-bearing, high-thoron-content sands of Kerala have a resident population of 200,000 living for generations at these high-background radiation levels [11]. The inhalational component in Kerala is smaller than the external gamma component, constituting only 30% of the total dose, whereas the external gamma dose comprises more than 50% of the total dose [11]. The frequency of micronuclei formation in newborns living in the HBRA of kerala was not different from newborns living in normal radiation areas [12].

C. L. L. Sanders, *Radiation Hormesis and the Linear-No-Threshold Assumption,*
DOI: 10.1007/978-3-642-03720-7_3, © Springer Verlag Berlin Heidelberg 2010

**Table 3.1** Origins of natural radioactivity in a 70-kg human body [2]

| Nuclide | Total mass | Disintegrations per day |
|---|---|---|
| Uranium isotopes | 90 μg | 95 thousand |
| Thorium isotopes | 30 μg | 9.5 thousand |
| Potassium-40 | 17 mg | 380 million |
| Radium isotopes | 31 pg | 95 thousand |
| Carbon-14 | 22 ng | 320 million |
| Tritium | 0.06 pg | 2 million |
| Polonium isotopes | 0.2 pg | 3.2 million |

**Table 3.2** Dose contributions of high-background radiation areas in the world [6]

| Geographic location | Annual dose (mSv) |
|---|---|
| Guarapari Beach, Brazil | Up to 790 |
| Ramsar, Iran | Up to 700 |
| SW France | Up to 88 |
| Kerala Beach, India | Up to 35 |
| Ataxa Beach | Up to 25 |
| Sweden | Up to 18 |
| U.S. Rocky Mountain States | 6–12 |
| Evacuated Land near Chernobyl | 6 |
| World Average | 2.4 |
| U.S. Gulf States | 0.8–1.2 |

**Table 3.3** Excess relative risk of common cancers in residents of Yangjiang, China [8]

| Site of cancer | ERR/Sv |
|---|---|
| Liver | −0.99 |
| Stomach | −0.27 |
| Lung | −0.68 |

There was no increase in all cancer mortality, all cause mortality, or decrease in longevity in native populations exposed to high levels of background radiation when compared with low-dose regions over an annual dose range of 0.8–700 mSv [13–15]. More rapid DNA repair was observed in persons living in Ramsar, Iran, who were exposed to high-background radiation [16]. Studies of Brazilian HBRAs showed both nonsignificant slightly increased and decreased cancer mortality [17]. No increase in cancer incidence has been found in inhabitants of Yangjiang, China [8, 18], Kerala, India [19], or in Ramsar, Iran [20–22]. A negative excess cancer risk was found in Yangjiang, China, at a natural background exposure of 9.9 mSv/year (Table 3.3) [23]. Lung cancer in Ramsar showed a

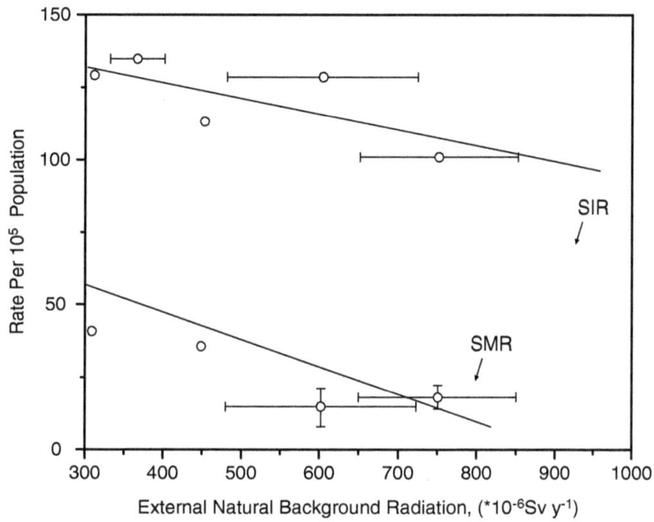


**Fig. 3.1** Cancer incidence and mortality in cities of India [25]

general negative correlation with natural radon levels [24]. A negative correlation was found for all cancer (SIR and SMR) with natural background dose in cities of India (Fig. 3.1) [25]. No increase in overall mortality or birth defects was found in case–control studies in Ramsar, Iran [26] or in Kerala, India [27].

Ramsar, a coastal city in northern Iran, has the highest level of background radiation level in an inhabited region of the world; the dose level is more than 100 mSv/year. This dose is more than five times higher than the 20 mSv/year permitted for radiation workers or 55–200 times more than the average global dose. There have been no ill effects to populations in Ramsar exposed to these high-radiation levels [5]. No significant cytogenetic effects in blood lymphocytes have been observed in Ramsar compared with those living in normal background levels. Physicians in Ramsar have not reported an increase in cancer rates [4, 21, 22]. An *in vitro* challenge dose of 1.5 Gy given to blood lymphocytes of inhabitants of Ramsar showed significantly reduced chromosomal damage compared with residents of normal background levels [22].

The age-adjusted cancer mortality rate for the U.S. population (1950–1967) decreased with increasing background radiation (Fig. 3.2). A 20% lower cancer mortality rate was found in Idaho, Colorado, and New Mexico than in Louisiana, Mississippi, and Alabama where background radiation levels were nearly five times less than for those living in the mountain states [28]. The incidence of leukemia and lymphoma was 19% less in males and 6% less in females for those living in the United States at an altitude of 2,000–5,300 feet when compared with those living at an altitude of <500 feet [29]. Correlation coefficients between mortality rates for various cancers and diseases of the heart and background dose levels in 43 urban populations are shown in Table 3.4. All correlations were negative with increasing background dose [30].

About 250,000 Americans receive background exposures of ~40 mSv/year living mostly in a handful of Rocky Mountain states, where lung cancer rates are much less than predicted by the EPA using the LNT assumption [31, 32]. Colorado indoor radon levels are

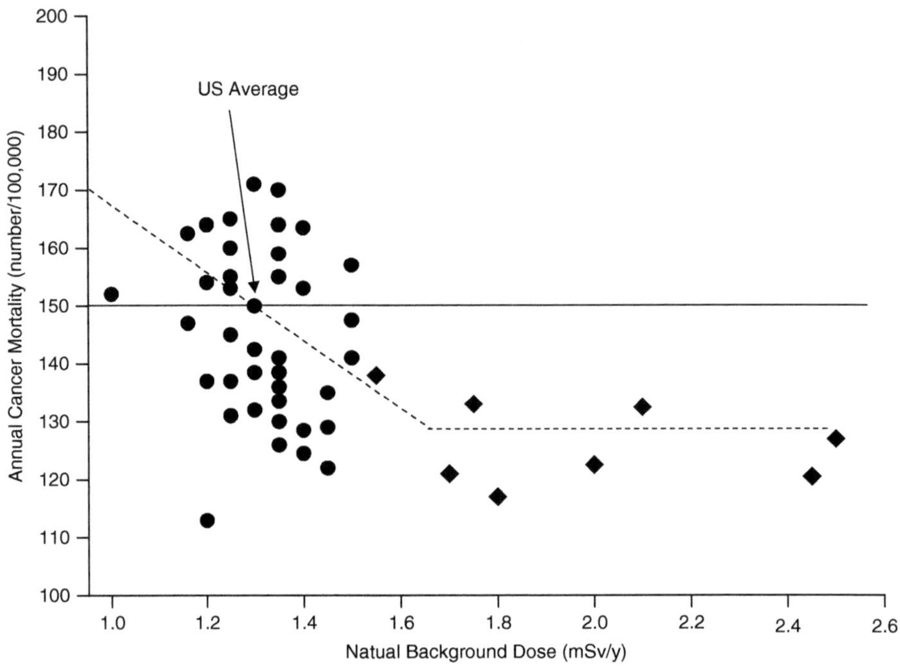


**Fig. 3.2** Cancer mortality in U.S. states [28]

**Table 3.4** Correlation coefficient of disease rate mortality and background radiation dose and level of statistical significance for cancer and heart disease in 43 urban populations of the United States (year 1960) [30]

| Cause of mortality | Correlation coefficient | Statistical significance |
|---|---|---|
| Cancer respiratory organs | −0.512 | $p < 0.001$ |
| Cancer buccal cavity and pharynx | −0.438 | $p < 0.01$ |
| Cancer digestive organs and peritoneum | −0.235 | $p < 0.10$ |
| All cancers | −0.300 | $p = 0.05$ |
| Heart disease | −0.257 | $p < 0.10$ |

well above the national average, averaging 7.3 pCi/L. The USEPA estimates the average indoor radon level nationwide is 1.3 pCi/L [33]. Relative to other states, Colorado has the third lowest lung cancer death rate in the nation. For the period 1993–1997, the Colorado cancer death rate per 100,000 population was 48.2 among males and 25.6 among females. These rates are well below the national averages of 69.4 for males and 34.0 for females.

Proponents of the LNT assumption claim that each disintegration from naturally occurring radiation is harmful, giving a calculable cancer risk per disintegration. Studies clearly show that this is not true. The risk of cancer is zero or less than zero at background radiation doses that are up to a hundred times the world annual average dose of 2.5 mSv.

Utilization of the LNT assumption to estimate cancer risk from exposure to background radiation is not credible [34]. The large preponderance of epidemiological data demonstrates no effect or a protection against cancer for those living in HBRA regions of the world [35–37].

## References


1. Cardis E (2005) Commentary on information that can be drawn from studies of areas with high levels of natural radiation. Int Congress Ser 1276:118–123
2. Wei L (1997) High background radiation area-an important source of exploring the health effects of low dose ionizing radiation. In: Wei L, Sugahara T, Tao A (eds) High levels of natural radiation, radiation dose and health effects. Elsevier, Amsterdam, The Netherlands, pp 1–14
3. Clouvas A, Xanthos S, Antonopoulos-Domis M (2006) Simultaneous measurements of indoor radon, radon-thoron progeny and high-resolution gamma spectrometry in Greek dwellings. Radiat Pro Dosimetry 118:482–490
4. Mortazavi SMJ, Ghiassi-Nejad M, Karam PA et al (2006) Cancer incidence in areas with elevated levels of natural radiation. Int J Low Radiat 2:20–27
5. Monfared AS, Jalail F, Sedaghat S et al (2006) High natural background radiation areas in Ramsar, Iran: Can inhabitants feel safe? Int J Low Radiat 3:171–177
6. Jaworowski Z (2001) Ionizing radiation in the 20th century and beyond. In: Symposium Entwicklungen im Strahleschutz, Munich, November 29. Available at http://www.cns-snc.ca/branches/Toronto/radiation/
7. UNSCEAR (2000) Sources and effects of ionizing radiation. United Nations Scientific Committee on the Effects of Atomic Radiation 2000 Report to the General Assembly, with Annexes. Volume II: Effects. No. E.00.IX.4. United Nations, New York, NY
8. Wei L, Sugahara T (2000) An introductory overview of the epidemiological study on the population at the high background radiation areas in Yangjiang, China. J Radiat Res (Tokyo) 41(Suppl):1–7
9. National Research Council, Committee to Assess Health Risks from Exposure to Low Levels of Ionizing Radiation (2005) Health risks from exposure to low levels of ionizing radiation: BEIR VII-Phase 2. National Academies Press, Washington, DC
10. Nair K, Nambi KSV, Amma NS et al (1999) Population study in the high natural background radiation area in Kerala, India. Radiat Res 152:S145–S148
11. Chougaonkar MP, Eappen KP, Ramachandran TV et al (2004) Profiles of doses to the population living in the high background radiation areas in Kerala. Indian J Nat Radioact 71:275–297
12. Birajalaxmi D, Karuppasamy CV (2009) Spontaneous frequency of micronuclei among the newborns from high level natural radiation areas of Kerala in the southwest coast of India. Int J Radiat Biol 85:272–280
13. National Council on Radiation Protection and Measurements (NCRP) (2001) Evaluation of the linear-non-threshold model for ionizing and measurements. NCRP, p 6
14. Luxin WEI, Sugahara T, Tao Z (1997) High level of natural radiation. Radiation dose and health effects. Elsevier, Amsterdam
15. Fornalski KW, Dobrzynski L (2010) The healthy worker effect and nuclear industry workers. Dose-Response (in press)
16. Masoomi JR, Mohammadi S, amino M, Ghiassi-Nejad M (2006) High background radiation areas of Ramsar in Iran: evaluation of DNA damage by alkaline single cell gel electrophoresis (SCGE). J Environ Radioact 86:176–186

17. Veiga LHS, Koifman S (2005) Pattern of cancer mortality in some Brazilian HBRAs. Int Congress Ser 1276:110–113
18. Zou J, Tao Z, Sun Q et al (2005) Cancer and non-cancer epidemiological study in the high background radiation area of Yangjiang, China. Int Congress Ser 1276:97–101
19. Ankathil R, Nair RK, Padmavathi J et al (2005) Review of studies in high level natural radiation areas in India. In: Proceedings of the 48th Annual Meeting of the Japan Radiation Research Society/the First Asian Congress of Radiation Research, Research Institute for Radiation Biology and Medicine, Hiroshima University, Japan. Abstract S4-1-1, p 79
20. Ghiassi-Nejad M, Beitollahi MM, Fallahian N et al (2005) New findings in the very high natural radiation area of Ramsar, Iran. Int Congress Ser 1276:13–16
21. Mortazavi SMJ, Ghiassi-Nejad M, Rezaiean M (2005) Cancer risk due to exposure to high levels of natural radon in the inhabitants of Ramsar, Iran. Int Congress Ser 1276:436–437
22. Mortazavi SMJ, Ikushima T (2006) Open questions regarding implications of radioadaptive response in the estimation of the risks of low-level exposures in nuclear workers. Int J Low Radiat 2:88–96
23. Sun Q et al (2000) Excess relative risk of solid cancer mortality after prolonged exposure to naturally occurring high background radiation of Yangjiang, China. J Radiat Res 41(Suppl):43–52
24. Mosavi-Jarrahi A, Akiba MMS, Yazdizadeh B et al (2005) Mortality and morbidity from cancer in the population exposed to high level of natural radiation area in Ramsar, Iran. Int Congress Ser 1276:106–109
25. Nambi KSV, Soman SD (1987) Environmental radiation and cancer in India. Health Phys 52:653–657
26. Monfared AS, Jalali F, Mozdarani H et al (2005) Living in high natural background radiation areas in Ramsar, Iran. Is it dangerous for health? Int Congress Ser 1276:438–439
27. Thampi MV, Cheriyan VD, Jaikrishan G et al (2005) Investigations on the health effects of human populations residing in the high-level natural radiation areas in Kerala in the southwest coast of India. Int Congress Ser 1276:8–12
28. Frigerio NA, Stowe RS (1975) Carcinogenic and genetic hazard from background radiation. In: Symposium on Biological Effects of Low-Level Radiation Pertinent to Protection of Man and His Environment, Chicago, IL, IAEA-SM-202/805, Vol. 2. International Atomic Energy Agency, Vienna, Austria, pp 385–393
29. Craig L, Seidman H (1961) Leukemia and lymphoma mortality in relation to cosmic radiation. Blood 17:319
30. Hickey RJ, Bowers EJ, Spence DE et al (1981) Low level ionizing radiation and human mortality: multi-regional epidemiological studies. Health Phys 40:625–641
31. American Cancer Society (1997) Cancer Facts & Figures-1997. American Cancer Society, 7Atlanta
32. Jagger J (1998) Natural background radiation and cancer death in Rocky Mountain states and Gulf Coast states. Health Phys 75:428–430
33. City of Fort Collins, CO (2005) Air quality department. http://www.ci.fort-collins.co.us/airqulaity/radon-health
34. Higson DJ, Boreham DR, Brooks AL et al (2007) Effects of low doses of radiation: joint statement from the following participants at the 15th Pacific Basin Nuclear Conference, sessions held in Sydney, Australia, Wednesday 18 October 2006. Dose Response 5:259–262
35. Thorne MC (2003) Background radiation: natural and man-made. J Radiol Prot 23:29–42
36. International Commission on Radiological Protection (1999) Protection of the public in situations of prolonged radiation exposure. ICRP Publication 82: Ann ICRP 29 (1–2)
37. USEPA (U.S. Environmental Protection Agency) (1993) Home buyers and sellers guide to radon. Report 402-R-93–003. Available at http://www.radon-levels.com

# Accidents, Tests, and Incidents 4

> The precautionary principle, an offspring of the LNT, leads to unacceptable societal penalties, as demonstrated in the aftermath of the Chernobyl catastrophe [1].

The U.S. radium dial painters of the 1920s comprised an early cohort of several thousand workers at increased risk of developing radiation induced cancers. Since the last year of World War II, there have been a variety of other nuclear events, including A-bombs dropped on Hiroshima and Nagasaki in 1945, thousands of underground, underwater, surface, and airborne nuclear weapon tests at various sites throughout the world, gross radioactive contamination of the Russian Mayak nuclear site and associated exposure of residents along the Techa River, the eastern Urals nuclear waste tank explosion in 1957, the Chernobyl nuclear reactor accident in 1986 in the Ukraine, and Taiwan residential buildings contaminated with $^{60}$Co. All these activities resulted in exposure of large human populations to low doses of ionizing radiation. These populations have been extensively studied in a myriad of epidemiological studies, none of which show increased cancer mortality at low doses, but instead show benefits with less than expected risks from low-dose radiation exposures (Table 4.1) [1–7].

## 4.1 Radium Dial Painters

One of the earliest populations exposed to ionizing radiation were the radium dial painters, who comprised several thousand women that spent many years tipping the end of radium-laden paintbrushes with their tongue before painting watch dials. The radium-laced paint contained both $^{226}$Ra and $^{228}$Ra. Studies of radium dial painters were started during World War II. Eventually, all the studies were concentrated at the Center for Human Radiobiology at the Argonne National Laboratory, Illinois. By the time the radium study was terminated in 1993, a total of 3,161 radium dial painters had been identified, of which 1,575 had been extensively studied [8].

High-dose dial painters were diagnosed with bone sarcomas and carcinomas of the paranasal sinuses or mastoid air cells. Radium-induced cancers were limited to high-dose radium dial painters who started work before 1926 [8]. There were 65 cases of bone cancer in those receiving bone doses >10 Gy and no cases in dial painters with bone doses <10 Gy,

C. L. L. Sanders, *Radiation Hormesis and the Linear-No-Threshold Assumption*,
DOI: 10.1007/978-3-642-03720-7_4, © Springer Verlag Berlin Heidelberg 2010

**Table 4.1** Cancer mortality among several irradiated human populations

| Event | Initial year | Population size | Radiation-induced cancer deaths |
|---|---|---|---|
| Japanese A-bombs | 1945 | 86,572 | 527[a] |
| USSR waste tank | 1957 | 7,852 | 0 |
| Mayak | 1948 | 6,293 | ~100[a] |
| Techa river | 1949 | 29,873 | 46[a] |
| Chernobyl | 1986 | ~6 million | 9 |

[a]Most deaths observed at high doses represent <10% of the exposed population

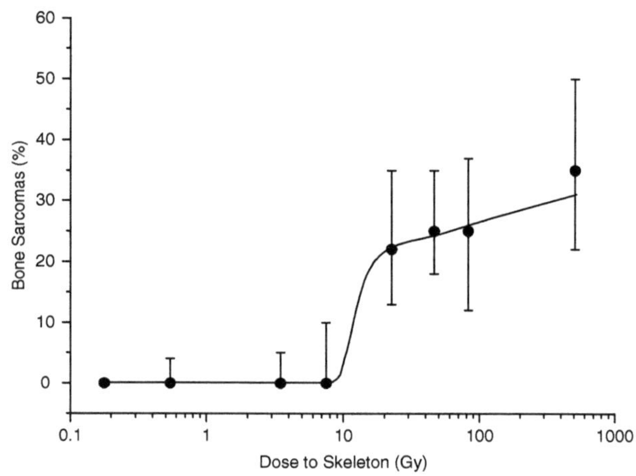


**Fig. 4.1** Incidence of bone cancer in radium dial painters as a function of skeletal dose [9]

providing unequivocal evidence for a very large threshold dose (Fig. 4.1) [9–12]. Lifespan survival time was greater than expected for both U.S. and U.K. radium dial painters whose skeletal doses were <10 Gy [13, 14].

$^{224}$Ra has a 3.6-day half-life, decaying by $\alpha$ and $\beta$ emissions. Injected $^{224}$Ra has been used in Germany to treat several thousand patients with bone tuberculosis and ankylosing spondylitis. Bone sarcomas are the main observation in patients receiving high-dose $^{224}$Ra treatment. Bone sarcomas occurred in 35 of 204 juveniles (11 Gy average bone dose) and in 13 of 612 adults (2 Gy average bone dose) [15]. A threshold of about 1 Gy was found for bone tumors in these German radium patients [10]. A similar threshold for bone cancer was also seen in dogs injected with radium and other alpha emitters [16].

## 4.2 Nuclear Weapons Tests

Thousands of nuclear weapon tests have been carried out since 1944 by the United States, Russia, England, France, China, Pakistan, and India. The latest test occurred in 2009 in North Korea. Nuclear weapon testing has added, at most, 5% to annual natural background radiation at the height of weapons testing; currently, it adds <0.2% [17]. A total of 21,357 U.K. and 210,000 U.S. servicemen participated in nuclear weapons tests. No evident trend toward more or less all cause or all cancer mortality was found in epidemiological studies where referents were matched servicemen who did not participate in nuclear tests [18, 19]. Participants in the U.S. Five Series Study found an all-cause-mortality SMR of 0.71 and an all-cancer-mortality SMR of 0.74 compared with rates in the U.S. general population [20].

Semipalatinsk is the name associated with the USSR atomic device testing site in northern Kazakhstan, just south of the Altai region of Siberia. From 1949 to 1989, the former USSR conducted more than 450 nuclear weapons tests at the Semipalatinsk site. Dolon village was ~110 km down-wind from Semipalatinsk. The mean cumulative radiation dose from nuclear fallout to residents of Dolon was estimated at 0.5 Gy [21]. The Dolon cohort was not adequately controlled for smoking. No cancer risk data were given for the lowest dose cohort (20–70 mSv). Several dose groups were combined for the next cohort (70–249 mSv). As a result, it was not possible to determine cancer risk at radiation doses <200 mSv. No dose–response trend was found for all cause or all cancer mortality [22].

## 4.3 Mayak and Techa River Residents

In 1948, a nuclear weapons production complex, Mayak, was established on the Techa River in the Southern Urals, approximately 100 km northeast of Chelyabinsk. A town of approximately 100,000 or more people (Chelyabinsk-65, now Ozersk) was constructed nearby to support and staff the facility. Inadequacies in technology and safety procedures resulted in massive releases of radioactivity into the surrounding environs as well as significant worker exposure during the first 5 years of operation. Radioactive material released from Mayak into the Techa River from 1948 to 1955 caused the surrounding areas to become heavily contaminated. High concentrations of $^{90}Sr$ and $^{137}Cs$ were incorporated into the local food and water supply. As a result, internal radionuclide doses reached very significant levels in the closest villages. Today, portions of the river bank still emit about 0.01 Gy/h [23].

The U.S. National Cancer Institute (NCI) has studied about 30,000 people who lived along the banks of the Techa River between 1950 and 1960. Several thousand children and adults received a mean external radiation dose of 100–200 mSv and a mean internal red marrow dose of 100–200 mSv [23, 24]. According to the NCI, only 46 cancer-related deaths resulted from excess radiation exposure [23].

## 4.4 Eastern Urals Nuclear Waste Tank Explosion

In 1957, an explosion in a nuclear waste storage tank in the Eastern Urals of Russia released a large plume of long-lived radionuclides covering an area to the northeast of some 25,000 km$^2$. The accident occurred at a facility involved in processing and storage of nuclear wastes. The waste tank had a volume of about 300 m$^3$. A chemical explosion with

**Table 4.2** Prevention of cancer in residents living downwind from the 1957 nuclear storage tank explosion in the Eastern Urals of Russia [25]

| Mean cumulative dose (mSv) | PROFAC, all solid cancer mortality |
|---|---|
| 40 | 27 |
| 120 | 39 |
| 496 | 28 |

an energy release of 5–10 tons of TNT aerosolized about 70–80 tons of wastes with a total activity of about 20 MCi. Of this, about 18 MCi was deposited in the immediate area of the explosion site, and about 2 MCi was dispersed over a large area downwind contaminating 22 villages. About 600 residents were evacuated about 7–10 days after the accident. About 10,000 people were evacuated by 18 months after the accident. According to Russian reports, maximum exposures reached 0.5 Gy. No acute radiation injuries were observed. Cancer mortality was significantly reduced in 7,852 inhabitants by 28, 39, and 27% for mean dose-cohorts of 490, 120, and 40 mGy, respectively (Table 4.2) [25].

## 4.5 Japanese A-Bomb Survivors

On August 6 and 9, 1945, the U.S. dropped atomic bombs on the Japanese cities of Hiroshima and Nagasaki, respectively. The United States and Japan launched a giant epidemiological study that included all residents of Hiroshima and Nagasaki who survived the atomic explosions within a 10-km radius. Clear evidence of radiation benefit has been shown in A-bomb survivors following more than 50 years of concentrated study by radiation scientists in a large survivor population with detailed long-term medical follow up, and reliable individual dose estimates. Neutron exposure may explain some of cancer risks observed at lower doses [26, 27].

Accelerated aging was noted by the early 1970s in Japanese A-bomb survivors who received radiation doses >1 Gy [28]. However, mortality rates in 120,321 atomic bomb survivors were not increased at doses <490 mSv [29]. Mortality was less than expected at lower doses with A-bomb survivors living longer than the nonexposed Japanese cohort [30, 31]. There was no evidence of an increased prevalence of birth defects, adult-onset hypertension, diabetes mellitus, hypercholesteremia, ischemic heart disease, and stroke in about 12,000 offspring of A-bomb survivors [32, 33].

Excess solid cancer mortality in A-bomb survivors (1950–1990) consisted of 334 deaths in 86,572 people for an incidence of 0.4%. Cancer incidence in 32,849 survivors receiving <100 mSv was 0.26% and 1.7% for those receiving >100 mSv. After 60 years of study, the results showed that 527 people died as a result of radiation received: 87 from leukemia and 440 from solid cancers. Most excess cancer deaths were found in those that received doses >1 Gy.

An increased incidence of leukemia and solid cancers in Japanese A-bomb survivors was only found in cohorts with doses >200 mSv. Solid cancer and leukemia incidences were less than expected at doses <200 mSv [26, 34–36]. There was no significant excess mortality at doses <500 mSv [29].

## 4.6
## Taiwan Contaminated Buildings

Recycled steel contaminated with cobalt-60 was used in the construction of more than 180 apartment buildings in Taiwan from 1982–1984, housing 10,000 persons for 9–20 years. Radioactivity was discovered in the first apartment in 1992 [37–39]. The average whole-body dose was 0.5 Sv given at an average dose rate of 50 mSv/year. About 1,100 persons received cumulative doses of 4 Sv from 1983 to 2003.

The LNT assumption used by the ICRP predicted 302 cancer deaths in the Taiwan study, 232 spontaneous, and 70 caused by radiation [38, 39]. However, few cases of fatal cancer were found in Taiwan building residents from 1983 to 2003. The cancer mortality rate of residents was 3.5 per 100,000 person-years, whereas the expected spontaneous cancer mortality rate was 116 persons per 100,000 person-years. Only seven cancer deaths were observed out of an expected 232 (RR = 0.030), which gave a PROFAC value of 0.97. The number of hereditary defects was also very low with three cases observed and 48 cases expected (RR = 0.063) [38, 39].

A follow-up study examining cancer incidence rather than cancer mortality failed to show the same dramatic protection, although confirming the presence of hormesis in both genders. When compared with the reference population, the study population had a lower incidence for all cancers combined and for all solid cancers combined (Table 4.3) [37]. Another follow-up study by the same research team showed only 117 cancers in a cohort of 6,242 apartment residents with a mortality much below the average cancer mortality [40]. One wonders if the comparatively high SIR for cancer compared with the SMR, even though the SIR values are much below expected values (Table 4.3), may be due to a "treatment effect" of low-dose radiation stimulating apoptosis of tumor cells.

## 4.7
## Chernobyl

On 26 April 1986, a reactor at the Chernobyl nuclear power plant complex was destroyed with a reactor core melt-down. A total of 134 Chernobyl employees developed acute radiation syndrome (ARS); of these, 28 died from acute radiation sickness and 2 from fire and falling debris; a further 19 died from 1987–2004. The liquidators comprised 240,000 people who received a cumulative mean dose of 100 mSv. Evacuees and residents in so-called low- and high-dose areas comprised nearly 5.5 million people with

**Table 4.3** Standardized incidence ratios (SIR) for exposed Taiwanese apartment dwellers, 1983–2002 [37]

| Cancer site | SIR | 95% CI |
|---|---|---|
| All cancers | 0.8 | 0.7–1.0 |
| Solid cancers | 0.7 | 0.6–0.9 |
| Liver | 0.6 | 0.3–1.2 |
| Lung | 0.8 | 0.4–1.5 |

**Table 4.4** Exposed populations to Chernobyl radioactive fallout and cumulative radiation doses

| Population | Number of people | Mean dose (mSv) |
|---|---|---|
| Liquidators | 240,000 | 100 |
| Evacuees | 116,000 | 33 |
| Low contaminated areas | 5,200,000 | 10 |
| High contaminated areas | 270,000 | 50 |

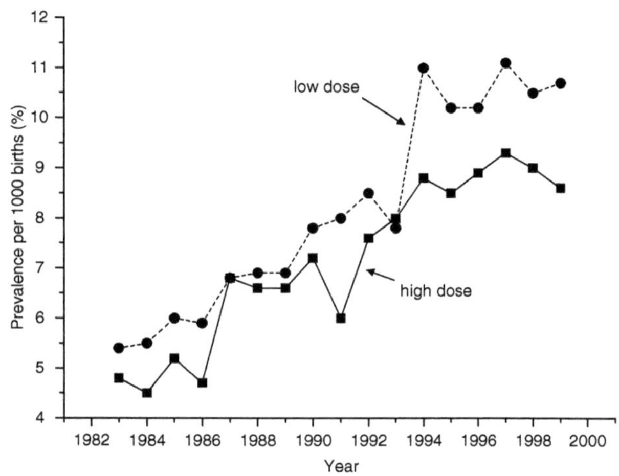


**Fig. 4.2** Prevalence at birth of nine types of congenital malformations in oblasts with high and low levels of radionuclide contamination from Chernobyl fallout [43]

cumulative doses of 10–50 mSv (Table 4.4). Low-LET, low-dose-rate radiation exposure from Chernobyl fallout was primarily due to $\gamma$ radiation from $^{137}Cs$, $^{90}Sr$, and radioiodines. Lung doses in some cohorts were as high as 0.6 Gy due mostly to inhalation of radionuclides [41].

The Chernobyl accident caused the displacement of large populations from regions of high radioactive deposition. The resultant anxiety associated with poor hygiene, poor nutrition, and an overall poor quality of life led to a disastrous morbidity. The fear of radiation hazards was so great in countries surrounding the Chernobyl accident site that more than 100,000 pregnant women had voluntary abortions performed [42]. In addition, more than 300,000 people from radionuclide-contaminated regions, where radiation doses were only a few times the natural background level, were needlessly evacuated leading to psychological, social, and economic problems [1]. This was fueled by the misleading concept that even the smallest dose of radiation is harmful.

Exposure *in utero* to high Chernobyl fallout radiation did not increase the prevalence of birth defects (Fig. 4.2), mental retardation, learning, hyperkinetic, or behavioral disorders [43, 44]. A similar irrational fear of radiation was seen following the Three Mile Island (TMI) accident. The highest rate of induced abortions for Pennsylvania women was 23/1,000 found in 1980 following the TMI accident [45]. The NRC estimated that the dose to the public from TMI averaged only 10 $\mu$Sv [46].

There was no increase in cancer incidence due to radiation in any exposed Chernobyl population [47]. No statistical evidence of increased leukemia incidence has been found in workers or populations associated with high radioactive fallout or exposure due to the Chernobyl accident [48].

An increase of thyroid cancers in children and adolescents was observed among residents living around Chernobyl. It was difficult to determine the cause of the increase in thyroid cancer after the accident because there was little screening for thyroid cancer before the accident, but extensive screening following the accident [47, 49]. A total of about 4,000 children and adolescents were estimated to have developed thyroid cancer, of which >99% were successfully treated. The SIR values for solid cancer among Chernobyl liquidators were greatly inflated by inclusion of cases of thyroid tumors, most of which were probably found from intensive postexposure screening; only nine of the thyroid cancer cases were fatal [50].

The LNT assumption, utilizing collective dose and ERR values for A-bomb survivor cancer risk, was used to estimate cancer mortality due to Chernobyl radiation. The calculated casualty figure was 53,400 excess cancer deaths over the following 50 years [2, 3]. Cardis, using the LNT assumption, later predicted a lifetime excess 9,000 cancer cases among the exposed Chernobyl cohorts [51].

Twenty years after the nuclear disaster, a three-volume report listed all recorded health effects. Only mortality from suicide was statistically increased [52]. The UN Chernobyl Forum (2006) found no increased cancer mortality during the first 20 years [43]. Consistent year-by-year evidence for radiation hormesis was seen in Chernobyl liquidators [53]. A cohort of 65,905 emergency workers who had received an external dose of 5–300 mSv showed annual SMRs for malignant neoplasms during 1991–1998 of 0.73–0.85 [54]. A group of 8,600 early cleanup workers who received external doses >50 mSv had 12% less cancer than expected [53]. In summary, mortality from all causes and all cancers has been consistently less than expected for every year examined since 1991 (Figs. 4.3 and 4.4).

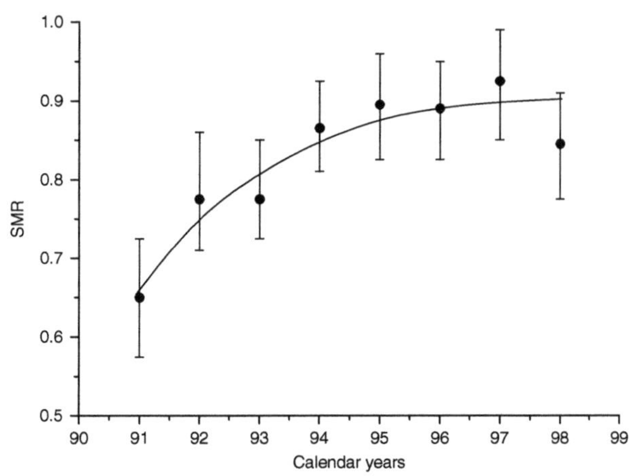


**Fig. 4.3** SMR for all cause mortality in Chernobyl liquidators [54]

**Fig. 4.4** SMR for all cancer mortality in Chernobyl liquidators [54]

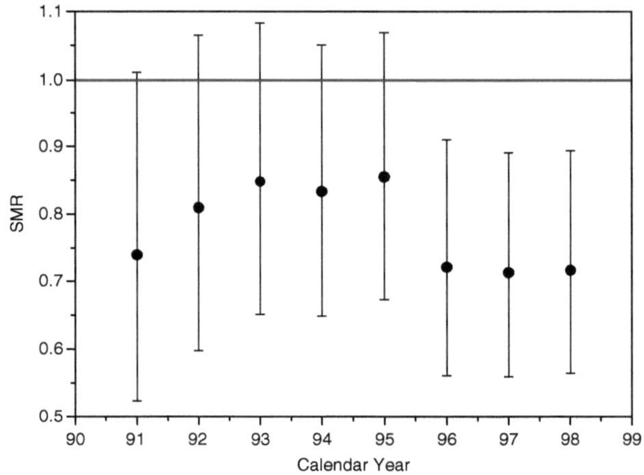


## References


1. Jaworowski Z (2006) Chernobyl: the fear of the unknown. Atomic Insight Guest Column. Available at http://www.atomicinsights.com/Guests/AGC_05–02–06.html
2. Jaworowski Z (1999) Radiation risks and ethics. Phys Today 52:24–29
3. Jaworowski Z (1998) Radiation risks in the 20th century: reality, illusions and ethics. Exec Intell Rev 25:15–19
4. Goldman M (1982) Ionizing radiation and its risks. West J Med 137:540–547
5. Goldman M, Filjushkin IV (1994) Low level radiation risks in people. Chin Med J 107:624–626
6. Anspaugh LR, Catlin RJ, Goldman M (1988) The global impact of the Chernobyl reactor accident. Science 242:1513–1519
7. Goldman M (1987) Chernobyl: a radiological perspective. Science 238:622–623
8. Rowland RE (1994) Radium in humans: a review of U. S. studies. Argonne National Laboratory, Argonne IL
9. Rowlands RE, Stheney AF, Lucas HF (1983) Dose-response relationships for radium-induced bone sarcomas. Health Phys 44(S1):15–31
10. Rowland RE (1997) Bone sarcoma in humans induced by radium: a threshold response? In: Radioprotection colloques, Proceedings of the Twenty-Seventh Annual Meeting of the European Society for Radiation Biology 32:C1/331–C1/338
11. Thomas RG (1994) The US radium luminisers: a case for a policy 'below regulatory concern'. J Radiat Prot 14:141–153
12. Evans RD (1974) Radium in man. Health Phys 27:497–510
13. Baverstock KF, Papworth D (1989) The UK radium luminizer survey. Br J Radiol 21:72–76
14. Stehney AF (1994) Survival times of pre-1950 U.S. women radium dial workers. In: Proceedings of the International Seminar Health effects of internally deposited radionuclides: emphasis on radium and thorium, Heidelberg, Germany, pp 149–155
15. Mays CW, Spiess H, Gerspach A (1978) Skeletal effects following $^{224}$Ra injections into humans. Health Phys 35:83–90
16. Raabe OG, Rosenblatt LS, Schlenker RA (1990) Interspecies scaling of risk for radium-induced bone cancer. Int J Radiat Biol 57:1047–1061
17. Thorne MC (2003) Background radiation: natural and man-made. J Radiol Prot 23:29–42

18. Muirhead CR, Bingham D, Haylock RGE et al (2003) Follow up of mortality and incidence of cancer 1952–98 in men from the UK who participated in the UK's atmospheric nuclear weapon tests and experimental programmes. Occup Environ Med 60:165–172
19. DTRA (2003) Radiation exposure is US atmospheric nuclear weapons testing. Nuclear Test Personnel Review Programme Defence Threat Reduction Agency Web Site: http://www.dtra.mil/news/fact/nw_ntprpre.html
20. Muirhead CR, Kendall GM, Darby SC et al (2004) Epidemiological studies of UK test veterans: II. Mortality and cancer incidence. J Radiol Prot 24:219–241
21. Imanaka T, Fukutani S, Yamamoto M et al (2005) External dose assessment for Dolon village due to fallouts from the semipalatinsk nuclear test site. In: Proceedings of the 48th Annual Meeting of the Japan Radiation Research Society/the First Asian Congress of Radiation Research, Research Institute for Radiation Biology and Medicine, Hiroshima University, Japan. Abstract W2–5, p 104
22. Bauer S, Gusev BI, Pivina LM et al (2005) Radiation exposure due to local fallout from Soviet atmospheric nuclear weapons testing in Kazakhstan: solid cancer mortality in the Semipalatinsk historical cohort, 1960–1999. Radiat Res 164:409–419
23. Goldman M (1997) The Russian radiation legacy: its integrated impact and lessons. Environ Health Perspect 105(Suppl 6):1385–1391
24. Kossenko MM (1996) Cancer mortality among Techa river residents and their offspring. Health Phys 71:77–82
25. Kostyuchenko VA, Krestina L Yu (1994) Long-term irradiation effects in the population evacuated from the East-Urals radioactive trace area. Sci Total Environ 142:19–125
26. Heidenreich W, Paretzke H, Jacob P (1997) No evidence for increased tumor rates below 200 mSv in the atomic bomb survivors data. Radiat Environ Biophys 36:205–207
27. Kellerer AM, Nekolla E (1997) Neutrons versus gamma ray risk estimates: inferences from the cancer incidence and mortality data in Hiroshima. Radiat Enviorn Biophys 36:73–83
28. Anderson RE, Key CR, Yamamoto T, Thorslund T (1974) Aging in Hiroshima and Nagasaki atomic bomb survivors. Am J Path 75:1–12
29. Cologne JB, Preston DL (2000) Longevity of atomic bomb survivors. Lancet 356:303–307
30. Mine M, Okumura Y, Ichimaru M et al (1990) Apparently beneficial effect of low to intermediate doses of A-bomb radiation on human lifespan. Int J Radiat Biol 58:1035–1043
31. Okajima S, Mine M, Nakamura T (1985) Mortality of registered A-bomb survivors in Nagasaki, Japan, 1970–1984. Radiat Res 103:419–431
32. Fujiwara S, Suyama A, Cologne JB et al (2008) Prevalence of adult-onset multifactorial disease among offspring of atomic bomb survivors. Radiat Res 170:451–457
33. Damilakis J (2004) Pregnancy and diagnostic X-rays. Eur Radiol Syllabus 14:33–39
34. Pierce DA, Shimizu Y, Preston DL et al (1996) Studies of the mortality of atomic bomb survivors. Report 12, Part 1. Cancer: 1950–1990. Radiat Res 146:1–27
35. Pierce DA, Preston DL (2000) Radiation-related cancer risks at low doses among atomic bomb survivors. Radiat Res 154:178–186
36. Little MP, Muirhead CR (1997) Evidence for curvilinearity in the cancer incidence dose-response in the Japanese atomic bomb survivors. Int J Radiat Biol 70:83–94
37. Hwang S-L, Guo H-R, Hsieh W-A et al (2006) Cancer risks in a population with prolonged low dose-rate $\gamma$-radiation exposure in radio-contaminated buildings, 1983–2002. Int J Radiat Biol 82:849–858
38. Chen WL, Luan YC, Shieh MC et al (2007) Effects of cobalt-60 exposure on health of Taiwan residents suggest new approach needed in radiation protection. Dose Response 5:63–75
39. Chen WL, Luan YC, Shieh MC et al (2004) Is chronic radiation an effective prophylaxis against cancer? J Am Physicians Surg 9:6–10
40. Hwang S-L et al (2008) Estimates of relative risks for cancers in a population after prolonged low-dose-rate radiation exposure: a follow-up assessment from 1983 to 2005. Radiat Res 170:143–148

41. Baverstock K, Williams D (2002) Chernobyl: an overlooked aspect? Science 299:44
42. Cigna AA, Durante M (2006) Radiation risks in normal and emergency situations. Springer, Dordrecht, The Netherlands, pp 49–67
43. World Health Organization (WHO) (2006) Health effects of the Chernobyl accident and special health care programmes. Report of the UN Chernobyl Forum expert group "health". Geneva, Switzerland
44. Igumnov SA (2007) Mental and behavioral disorders in Belarusian persons exposed *in utero* to radiation following the Chernobyl accident. In: Sixht LOWRAD Conference, Budapest, Hungary. Abstract, p 62
45. Pennsylvania past and present: Pennsylvania maturity, (1945–2003) The Pennsylvania manual, section 1. Pennsylvania Department of General Services, Harrisburg, PA
46. Nuclear Regulatory Commission (2005) Fact sheet on the accident at Three Mile Island. Nuclear Regulatory Commission, Revised. http://www.nrc.gov/reading-rm/doc-collections/fact-sheets/3mile-isle.html
47. Balonov M (2007) Third annual Warren K. Sinclair keynote address: retrospective analysis of impacts of the Chernobyl accident. Health Phys 93:383–409
48. Howe GR (2007) Leukemia following the Chernobyl accident. Health Phys 93:512–515
49. UNSCEAR (2000) United Nations Scientific Committee on the Effects of Atomic Radiation. Sources and effects of ionizing radiation. 2000 report to the general assembly, with annexes. Annex J, Volume II. United Nations, New York, pp 451–566
50. Ivanov VK (2007) Late cancer and non-cancer risks among Chernobyl emergency workers of Russia. Health Phys 93:470–479
51. Cardis E, Anspaugh L, Ivanov VK et al (1996) Estimated long term health effects of the Chernobyl accident. One decade after Chernobyl: summing up the consequences of the accident. International Atomic energy Agency, Vienna, pp 241–279
52. Rahu M, Tekkel M, Veidebaum T et al (1997) The Estonian study of Chernobyl cleanup workers II. Incidence of cancer and mortality. Radiat Res 147:641–652
53. Ivanov V, Iiyin L, Gorski A et al (2004) Radiation and epidemiological analysis for solid cancer incidence among nuclear workers who participated in recovery operations following the accident at the Chernobyl NPP. J Radiat Res (Tokyo) 45:41–44
54. Ivanov VK, Gorski AI, Maksioutov MA et al (2001) Mortality among the Chernobyl emergency workers: estimation of radiation risks (preliminary analysis). Health Phys 81:514–521

# Medical Exposures and Workers 5

The NCRP could turn on us at any time and come up with new recommendations, perhaps under pressure to implement the ICRP model. It is also clear from history that you can never relax your vigilance with so many groups trying to meddle and regulate in this arena

(Bob Dixon)

## 5.1 Radiotherapy for Noncancer Conditions

Relatively low dose radiotherapy has been used to treat a variety of acute and chronic inflammatory diseases and painful disorders for over 70 years [1]. These include nasopharyngeal radium radiotherapy for adenoid hypertrophy, and low dose X-ray radiotherapy for skin hemangioma, thymus gland enlargement, benign breast disease, and fertility problems. Recommended single doses were 0.3–1.0 Gy in 4–5 fractions for acute and 1–3 fractions for chronic diseases per week for total doses of 3–5 and 12 Gy, respectively [2, 3]. Decreased cancer risks among these exposed clinical groups will be examined in subsequent chapters.

Radon has been therapeutically used for centuries to treat infectious and inflammatory conditions, and as an analgesic and anti-inflammatory agent for arthritis, rheumatism, fibromyalgia, psoriasis, asthma and bronchitis [4–6]. Radon is helpful in the treatment and prevention of osteoarthritis [7]. Tens of thousands of people annually expose themselves to high levels of radon for therapeutic benefit in old mines, spas and clinics exposed by inhalation, bath or steam (Chap. 14). $^{222}$Rn gas enhances the activity of superoxide dismutase and catalase, inhibits lipid peroxidation, and enhances immune function (mitogen response, CD4 and CD8 positive cells, which are markers for helper T cells and killer T cells, respectively) [8, 9]. Epidemiological studies of radon therapy patients have not demonstrated any carcinogenic effect from radon therapy up to 1,000 Bq/m$^3$, or seven times the EPA limit (Chap. 8) [9].

C. L. L. Sanders, *Radiation Hormesis and the Linear-No-Threshold Assumption*,
DOI: 10.1007/978-3-642-03720-7_5, © Springer Verlag Berlin Heidelberg 2010

## 5.2 Diagnostic Radiation Exposures

Millions of diagnostic examinations are performed in the world every day. About 330 million people are exposed each year to low doses of radiation from diagnostic exposures, while about 5 million patients undergo radiotherapy at high doses, mostly for treatment of cancer [10]. Diagnostic radiation exposures from a large variety of procedures deliver radiation doses to patients that typically range from 0.02 to 300 mSv (Table 5.1). More than 60 million computed tomography (CT) scans are annually performed in the United States, including about 4 million in children. The time required for an average scan has decreased to less than 1 s. Typical effective doses from CT scans are about 10 mSv. This is similar to doses delivered to the majority of Japanese A-bomb survivors [12] and nuclear workers [13]. Using the LNT assumption, Brenner and Hall estimate that a 1–2% excess cancer risk can be attributed to CT scans by applying the LNT assumption [14]. Another study of patients receiving multiple CT examinations reported that "this patient subgroup may have a heightened risk of developing cancer from cumulative CT radiation exposure" [15]. A similar excess cancer risk was estimated by Cardis et al. for nuclear workers using the LNT assumption [13]. However, low doses associated with X-ray diagnostic examinations do not increase the incidence of leukemia or solid cancers, but are likely to reduce cancer risk [16–19].

Two epidemiological studies have examined cancer risk in 77,000 tuberculosis patients given multiple X-ray fluoroscopies [20–22]. The average lung dose in the Massachusetts study was 0.84 Gy and the RR for lung cancer was 0.8 (95% CI 0.6–1.0) [20]. The Canadian cohort of fluoroscopy patients had lung doses up to 3 Gy; the RR for lung cancer was <1.0 at lung doses <1.0 Gy [21]. Relative risks for lung cancer in fluoroscopy patients and in the contralateral lung in patients irradiated for breast cancer are plotted in Fig. 5.1. The threshold for increased relative risk of lung cancer was 1.5 Gy. Excess leukemia and lung cancer were not observed in girls receiving fluoroscopic examinations or cardiac catheterization [17].

**Table 5.1** Estimated range of radiation doses from diagnostic radiation exposures [11]

| Procedure | Comments | Dose range (mSv) |
|---|---|---|
| Conventional | Simple X-rays for bones, skull, abdomen | 0.02–10 |
| Conventional | Complex X-rays for GI series, barium enema, IVP | 3–10 |
| CT | Head, whole-body | 5–15 |
| Spiral CT | Head, whole-body | 10–20 |
| Angiography | Heart, aorta, peripheral, carotid, abdominal | 10–200 |
| Interventional | Stent placement, percutaneous dilatation, closures, biopsy | 10–300 |
| Internal emitters | Radioisotope studies | 3–14 |

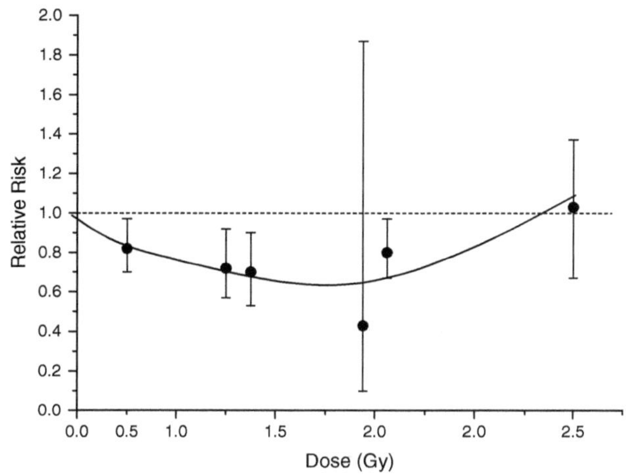


**Fig. 5.1** Relative risk of lung cancer in patients given repeated fluoroscopic examinations during therapy for tuberculosis [20, 21], and in contralateral lung lobe receiving radiation from radiotherapy for breast cancer [23]. The figure is redrawn from Rossi and Zaider [24]

## 5.3 Prenatal Exposures

Studies of Japanese A-bomb survivors exposed *in utero* have not shown a prenatal carcinogenic effect, either for leukemia or solid tumor formation. Similarly animal experiments have failed to show that *in utero* radiation exposure increases tumor formation at low doses. The risk of cancer from diagnostic X-rays given to the fetus or during early childhood has been investigated in many studies. Doll and Wakeford claim that prenatal X-ray examinations delivering 5–10 mSv was associated with an increase in childhood solid cancers and leukemia [25]. However, nineteen subsequent case–control studies and six cohort studies failed to demonstrate any significant cancer risk from pre- and postnatal X-ray exposures [26].

## 5.4 Radioiodine Therapy

Radioiodine has been used in the treatment of hyperthyroidism for more than 50 years. Cancer incidence was determined among 7,417 patients with hyperthyroidism treated with $^{131}I$ in the UK. The dose to the thyroid gland was very high, while the whole body dose from $^{131}I$ was 280 mSv. The relative risk from all cancer was significantly reduced, more in men (SIR 0.65) than in women (SIR 0.85), due in part to a significant decrease in lung cancer (Table 5.2) [27]. As expected, thyroid cancer was significantly increased. These findings were similar to a previous study at Massachusetts General Hospital, where a non-significant decrease in cancer mortality (SMR 0.9 [95% CI 0.7–1.1]) was found [28]. These studies show that cancer incidence is reduced in "unhealthy" populations.

**Table 5.2** Cancer incidence among 7,417 patients with hyperthyroidism treated with $^{131}I$ in the UK [27]

| Cancer site | SMR | 95% CI |
|---|---|---|
| All cancer | 0.90[a] | 0.82, 0.98 |
| Lip, oral cavity, pharynx | 0.37 | 0.09, 1.48 |
| Digestive organs | 0.96 | 0.82, 1.12 |
| Respiratory tract | 0.76[a] | 0.61, 0.95 |
| Breast | 0.93 | 0.74, 1.18 |
| Genitourinary tract | 0.80 | 0.62, 1.02 |
| Urinary bladder | 0.49 | 0.23, 1.03 |
| Brain | 1.20 | 0.60, 2.40 |
| Lymphatic and hematopoietic | 0.84 | 0.57, 1.24 |
| Lymphoma | 0.64 | 0.31, 1.34 |
| Leukemia | 1.17 | 0.68, 2.01 |
| Thyroid | 2.78[a] | 1.16, 6.67 |

[a] $p < 0.05$

## 5.5 Second Tumors in Radiotherapy Patients Treated for a Primary Tumor

Second primary tumors occurring after radiotherapy for the first primary tumor has become a major concern. Cumulative doses outside the tumor volume region range from a few mSv up to 60 Gy. Second tumors are mostly observed in tissues with absorbed doses >2 Gy of fractionated irradiation with increasing incidence as the dose increases. Lowering the dose-rate reduces carcinogenesis. Exceptions are thyroid cancers in children and breast cancers in adults as second cancers at doses as low as 100 mSv. Lung cancers in adults as second cancers were seen at lung doses as low as 500 mSv probably due to an interaction with cigarette smoking [29].

Over one million patients are treated each year in industrialized countries by radiotherapy. Approximately 50% of cancer patients in the U.S. are treated by radiotherapy, most fractionally exposed to high cumulative doses of 40–60 Gy in the tumor targeted region. Estimates of cancer risk to tissues adjacent to high-dose radiotherapy sites provide important data on risk of cancer from high to low dose exposures[30–36]. A pooled analysis of 11 studies showed that radiation treatment for cancer increased the risk of second primary cancers by ~10% over a dose range of 1–45 Gy [37–40]. However, there was great heterogeneity in risk of second tumors within organs, causing uncertainty in quantification of risk factors [40]. One study of radiotherapy for breast cancer reported a threshold of 0.6 Gy delivered in 30 fractions for second cancer formation [41].

High doses induce second cancers in organs near the "targeted" tumor while low doses in the same patient may reduce cancer risk in "distant" low dose exposed organs. Radiation dose decreases with distance from the tumor, such that many tissues will receive doses of <100 mSv [30]. Evidence of radiation benefit (less cancers than is expected in a patient not

**Table 5.3** Second primary tumors in patients receiving radiotherapy for prostate cancer [40, 42, 43]

| Dose (Gy) | Secondary site | RT/GP |
|---|---|---|
| 60 | Bladder | 1.09 (1.00–1.19)[a] |
| 60 | Rectum | 0.97 (0.68–1.37) |
| 24 | Colon | 1.06 (0.80–1.40) |
| 2 | Stomach | 0.81 (0.70–0.95) |
| 2 | Kidney | 1.00 (0.87–1.16) |
| 2 | Pancreas | 0.92 (0.79–1.07) |
| 0.3 | Lung | 0.82 (0.78–0.88) |
| 0.3 | Esophagus | 0.85 (0.68–1.06) |
| NA | All eight sites | 0.94 (0.85–1.05) |
| NA | All sites | 0.94 (0.84–1.06) |

[a]95% Confidence interval; *RT* radiotherapy; *GP* general population

receiving radiotherapy) were seen in patients receiving radiotherapy for cervical and prostate cancer (Table 5.3) [40]. In a series of three studies, the $RR_{RT/NonRT}$ values for breast carcinoma in patients receiving radiotherapy for carcinoma of the cervix were 0.78, 0.78 and 0.92; the dose to the breast was ~300 mSv [44–46].

Radiotherapy for breast cancer exposes the ipsilateral lung to a high dose while the contralateral lung receives less radiation. In one series of breast cancer patients the average dose to the contralateral lung was 4.6 Gy [47]. In another series of breast cancer patients treated with radiotherapy, the mean lung dose ranged from 9.9 to 15.5 Gy in the ipsilateral lung and 0.55–2.2 Gy in the contralateral lung [23]. The relative risk for lung cancer was 5.3 for the ipsilateral lung but only 0.48 for the contralateral lung [23].

## 5.6 Medical Workers

There are worldwide over 2 million persons involved in medical, dental and veterinarian radiological practices. Eight major studies of medical radiation workers have been carried out [48]: U.S. radiologists [49], UK radiologists [50], U.S. technologists [51], U.S. Army technologists [52], Chinese X-ray worker [52], Danish radiation therapy workers [53], Japanese technologists [54], and Canadian radiation workers [55, 56].

Early radiation dose estimates and exposure standards used the skin erythema dose, but became more precise with time after the implementation of the roentgen dose unit in 1928. Exposure of radiologists and radiologic technicians progressively decreased with time as stricter exposures were implemented by radiological organizations (Table 5.4).

In spite of high dose exposures, UK radiologists (1897–1920) had a noncancer SMR of 0.86 compared with all other male physicians. Noncancer mortality made up ~80% of all mortality. Post-1955 radiologists had an all cause SMR of 0.68 compared with nonradiologist,

**Table 5.4** History of radiation exposure among radiologists and technicians [48]

| Date | Dose |
|---|---|
| 1902 | 100 mSv/day based on fogging of a photographic plate |
| ~1910–1920 | 60 mSv/month [1/10 erythema] |
| ~1925 | 6 mSv/month [1/100 erythema dose] |
| 1928 | 300 mSv/year [using new roentgen unit] |
| 1940s | 100 mSv/year |
| 1950s | 50–150 mSv/year |
| 1957 | 50 mSv/year [ICRP limit] |
| 1990 | 20 mSv/year [ICRP limit] |

**Table 5.5** Influence of time period working as measure of radiation dose exposure on cancer mortality in British radiologists. The referent group was UK male nonradiologist physicians [50]

| | SMR cancer site | | | | |
|---|---|---|---|---|---|
| Date[a] | All cancer | Skin | Colon | Lung | Leukemia |
| 1897–1920 | 1.75 | 4.35 | 1.27 | 2.46 | 2.50 |
| 1921–1935 | 1.24 | 4.55 | 0.84 | 1.06 | 2.70 |
| 1936–1954 | 1.12 | 0.00 | 0.51 | 0.74 | 1.75 |
| 1955–1979 | 0.71 | 0.00 | 0.63 | 0.00 | 1.16 |

[a]Years joined British radiological societies

male physicians [50]. The doses for each of four time cohorts were estimated by Simon et al. [57] (Table 5.4). Post-1955 UK radiologists experienced a 32% lower SMR ($p < 0.001$) for deaths from all causes than that of all other physicians, and an SMR for noncancer deaths that was 36% lower ($p < 0.001$) than that of all UK physicians (Table 5.5) [58].

A hormetic effect was observed among 142,517 U.S. radiologic technologists, with SMRs for all cause and all cancer mortality of 0.69 and 0.79, respectively. Decreased risk was seen for all but one cancer type (Table 5.6) [59]. Mortality patterns were generally similar for male and female technologists. Benefits of radiation exposure were somewhat decreased with increasing years of certified work experience (Table 5.7) [59].

Among the 191,333 Canadian workers in the National Dose Registry, 57% were dental-medical workers, 31% were industrial workers and 12% were nuclear workers [56]. Medical workers experienced similar benefits from exposure to low dose radiation as did industrial and nuclear workers (Table 5.8) [55, 56].

A total of 1,416 Latvian female radiologists, radiological technicians and nurses exhibited an SIR for all cancer of 0.76 (95% CI: 0.59–0.96). The SIR for persons employed during/after 1972 was 0.67 (95% CI: 0.45–0.97). Less benefit was seen in women employed after age 40 [60].

**Table 5.6** Cancer mortality among US radiologic technologists [59]

| Cancer site | SMR | 95% CI |
|---|---|---|
| All cause | 0.69 | 0.68–0.71 |
| All cancer | 0.79 | 0.76–0.83 |
| Oral cavity & pharynx | 0.67 | 0.45–0.95 |
| Esophagus | 0.67 | 0.43–1.01 |
| Stomach | 0.67 | 0.50–0.88 |
| Colon | 0.78 | 0.67–0.90 |
| Rectum | 0.78 | 0.56–1.06 |
| Liver, gallbladder, bile duct | 0.73 | 0.51–1.02 |
| Pancreas | 0.83 | 0.67–1.02 |
| Larynx | 0.49 | 0.21–0.96 |
| Lung, trachea, bronchus | 0.74 | 0.67–0.82 |
| Bone | 0.39 | 0.11–1.01 |
| Skin, including melanoma | 0.62 | 0.44–0.84 |
| Breast | 0.99 | 0.90–1.09 |
| Cervix | 0.27 | 0.16–0.39 |
| Uterus | 0.87 | 0.59–1.13 |
| Prostate | 0.79 | 0.56–1.08 |
| Bladder | 0.91 | 0.63–1.28 |
| CNS | 0.83 | 0.65–1.03 |
| Thyroid | 0.60 | 0.16–1.50 |
| Lympho-reticulum cell sarcoma | 1.03 | 0.72–1.38 |
| Hodgkin's lymphoma | 0.87 | 0.58–1.26 |
| Multiple myeloma | 0.71 | 0.45–1.06 |
| Leukemia | 0.93 | 0.76–1.13 |

**Table 5.7** SMR for mortality in U.S. radiologic technologists by number of years certified [59]

| Cause of death | Number of years certified | | | |
|---|---|---|---|---|
| | <10 | 10–19 | 20–29 | 30+ |
| All causes | 0.6[a] | 0.7[a] | 0.8[a] | 0.8[a] |
| All cancers | 0.7[a] | 0.7[a] | 0.9[a] | 0.9[a] |
| Lung cancer | 0.7[a] | 0.7[a] | 0.8[a] | 0.8[a] |

[a] $p < 0.05$

In summary, exposure of large medical worker and patient populations to ionizing radiation from medical-dental procedures for noncancer conditions, to a wide variety of radiological diagnostic tests, to radioiodine therapy, to radiotherapy for cancer and to radiologists and radiological technologists results in less mortality from cancer and other diseases.

**Table 5.8** PROFAC X 100 (%) for cancer in two Canadian cohorts comprising male and male and female nuclear dental, medical, industrial and nuclear workers [55, 56]. The referent group is the general Canadian population

| Cancer incidence | Nuclear power workers (SMR) [55] | Dental, medical, industrial and nuclear workers (SIR) [56] |
|---|---|---|
| All cause | 43 | NA |
| All cancer | 32 | 21 |
| Leukemia | 23 | 28 |
| Non-Hodgkin's lymphoma | 13 | 29 |
| Esophagus | 27 | 48 |
| Colon | 15 | 10 |
| Rectal | 34 | 27 |
| Pancreas | 20 | 25 |
| Lung | 39 | 34 |
| Central nervous system | 27 | 21 |

## References

1. von Pannewitz G (1933) Die Rontgentherapie der Arthritis deformans. Ergebnisse der medizinischen Strahlenforschung 6:62–126
2. Seegenschmiedt MH, Micke O, Willich N (2004) Radiation therapy for nonmalignant diseases in Germany. Current concepts and future perspectives. Stranhlentherapie Onkologie 180:P718–P730
3. Rodel F, Keilholz L, Herrmann M et al (2007) Radiobiological mechanisms in inflammatory diseases of low-dose radiation therapy. Int J Radiat Biol 83:357–366
4. Cuttler JM (2004) Low-dose irradiation therapy to cure gas gangrene infections. Int J Low Radiat 1:318–328
5. Franke A, Reiner L, Pratzel HG et al (2000) Long-term efficacy of radon spa therapy in rheumatoid arthritis-a randomized, sham-controlled study and follow-up. Rheumatology 39: 894–902
6. Mitsunobu F, Yamaoka K, Hanamoto K et al (2003) Elevation of antioxidant enzymes in the clinical effects of radon and thermal therapy for bronchial asthma. J Radiat Res 44:95–99
7. Yamaoka K, Mitsunobu F, Hanamoto K et al (2005) Effects of radon and thermal therapy on osteoarthritis. Int Congress Ser 1276:249–250
8. Falkenbach A, Wolter N (1997) Radonthermalstollen-Kur zur Behandlung des Morbius Bechterew. Res Complementary Med 1997:277–283
9. Kant K (2007) Is exposure to over ground radon as dangerous as they say? In: Sixth LOWRAD Conference, Budapest, Hungary, Abstract, p 67
10. UNSCEAR (2000) Sources and effects of ionizing radiation. UNSCEAR 2000 Report to the General Assembly, with Scientific Annexes. United Nations, New York, p 1220
11. Committee to Assess Health Risks from Exposure to Low Levels of Ionizing Radiation (2005) BEIR VII – phase 2, health risks from exposure to low levels of ionizing radiation, National Research Council (National Academy of Sciences), Washington
12. Preston DL, Pierce DA, Shimizu Y et al (2004) Effect of recent changes in atomic bomb survivor dosimetry on cancer mortality risk estimates. Radiat Res 162:377–389

13. Cardis E, Vrijheid M, Blettner M et al (2007) The 15-country collaborative study of cancer risk among radiation workers in the nuclear industry: estimates of radiation-related cancer risks. Radiat Res 167:396–416
14. Brenner DJ, Hall EJ (2007) Computed tomography – an increasing source of radiation exposure. N Engl J Med 357:2277–2284
15. Griffey RT, Sodickson A (2009) Cumulative radiation exposure and cancer risk estimates in emergency department patients undergoing repeat or multiple CT. AJR 192:887–892
16. Scott BR, Sanders CL, Mitchel REJ, Boheham DR (2008) CT scans may reduce rather than increase the risk of cancer. J Am Physicians Surg 13:8–11
17. Kleinerman RA (2006) Cancer risks following diagnosis and therapeutic radiation exposure in children. Pediatr Radiol 36(Suppl 14):121–125
18. Linos A, Gray JE, Orvis AL et al (1980) Low dose radiation and leukemia. N Engl J Med 302:1101–1115
19. Boice JD, Morin MM, Glass AG et al (1991) Diagnostic X-ray procedures and risk of leukemia, lymphoma and multiple myeloma. JAMA 265:1290–1294
20. Davis FG, Boice JD, Hrubec Z et al (1989) Cancer mortality in a radiation-exposed cohort of Massachusetts tuberculosis patients. Cancer Res 49:6130–6136
21. Howe GR (1995) Lung cancer mortality between 1950 and 1987 after exposure to fractionated moderate-dose-rate ionizing radiation in the Canadian fluoroscopy chart study and a comparison with lung cancer mortality in the atomic bomb survivor study. Radiat Res 142: 295–305
22. Boice JD, Preston D, Davis FG, Monson RR (1991) Frequent chest X-ray fluoroscopy and breast cancer incidence among tuberculosis patients in Massachusetts. Radiat Res 125: 214–222
23. Neugut AI, Robinson E, Lee WC et al (1993) Lung cancer after radiation therapy for breast cancer. Cancer 71:3054–3057
24. Rossi HH, Zaider M (1997) Radiogenic lung cancer: the effects of low doses of low linear energy transfer (LET) radiation. Radiat Environ Biophys 36:85–88
25. Doll R, Wakeford R (1997) Risk of childhood cancer from fetal irradiation. Br J Radiol 70:130–139
26. Schulze-Rath R, Hammer GP, Blettner M (2008) Are pre- or postnatal diagnostic X-rays a risk factor for childhood cancer? A systematic review. Radiat Environ Biophys 47:301–312
27. Franklyn JA, Maisonneuve P, Sheppard M et al (1999) Cancer incidence and mortality after radioiodine treatment for hyperthyroidism: a population-based cohort study. Lancet 353: 2111–2115
28. Goldman MB, Maloof F, Monson RR et al (1988) Radioactive iodine therapy and breast cancer. Am J Epidemiol 127:969–980
29. Tubiana M (2009) Can we reduce the incidence of second primary malignancies occurring after radiotherapy? A critical review. Radiother Oncol 91:4–15
30. Hanford JM, Quimby EH, Frantz VK (1962) Cancer arising many years after radiation therapy. J Am Med Assoc 181:132–138
31. Schneider U, Walsh L (2008) Cancer risk estimates from the combined Japanese A-bomb and Hodgkin cohorts for doses relevant to radiotherapy. Radiat Environ Biophys 47:253–263
32. Pifer JW, Toyooka ET, Murray RW et al (1963) Neoplasms in children treated with X rays for thymic enalrgement. I. Neoplasms and mortality. J Natl Cancer Inst 31:1333–1356
33. Hempelmann LH, Hall WJ, Phillips M et al (1975) Neoplasms in persons treated with X-rays in infancy: fourth survey in 20 years. J Natl Cancer Inst 55:519–530
34. Hazen RW, Pifer JW, Toyooka ET et al (1966) Neoplasms following irradiation of the head. Cancer Res 26:305–311
35. Modan B, Bidatz D, Mart H (1974) Radiation-induced head and neck tumors. Lancet 1: 277–279

36. Shore RE, Albert RE, Pasternak BS (1976) Follow-up study of patients treated by X-ray epilation for tinea capitis. Arch Environ Health 31:17–24
37. Doherty MA, Rodgers A, Langlands AO (1986) Sarcoma of bone following therapeutic irradiation for breast carcinoma. Intern J Radiat Oncol Biol Phys 12:103
38. Thomas P et al (1990) Cancer recurrence after resection: T1 N0 non-small cell lung cancer. Ann Thorac Surg 49:242–247
39. Pastorino U (1994) Results of the Euroscan trial. Lung Cancer 11:94–95
40. Suit H, Goldberg S, Niemierko A et al (2007) Secondary carcinogenesis in patients treated with radiation: a review of data on radiation-induced cancers in humans, non-human primate, canine and rodent subjects. Radiat Res 167:12–42
41. Rubino C, de Vathaire F, Shamsaldin A et al (2003) Radiation dose, chemotherapy, hormonal treatment and risk of second cancer after breast cancer treatment. Br J Cancer 89:840–846
42. Brenner D, Curtis R, Hall E, Ron E (2000) Second malignancies in prostate carcinoa patients after radiotherapy as compared with surgery. Cancer 88:398–406
43. Pickles T, Phillips N (2002) The risk of second malignancy in men with prostate cancer treated with or without radiation in British Columbia, 1984–2000. Radiother Oncol 65:145–151
44. Kleinerman R, Boice J, Storm H et al (1995) Second primary cancer after treatment for cervical cancer. An International Cancer Registries Study. Cancer 76:442–452
45. Storm H (1988) Second primary cancer treatment for cervical cancer. Cancer 61:679–688
46. Arai T, Nakano T, Fukuhisa K et al (1991) Second cancer after radiation therapy for cancer of the uterine cervix. Cancer 67:398–405
47. Inskip PD, Stovall M, Flannery JT (1994) Lung cancer risk and radiation dose among women treated for breast cancer. J Natl Cancer Inst 86:983–988
48. Yoshinaga S, Mabuchi K, Sigurdson AJ et al (2004) Cancer risks among radiologists and radiologic technologists: review of epidemiologic studies. Radiology 233:313–321
49. Matanoski GM, Sternberg A, Elliott EA (1987) Dose radiation exposure produce a protective effect among radiologists? Health Phys 52:637–643
50. Berrington A, Darby SC, Weiss HA, Doll R (2001) 100 years of observation on British radiologists: mortality from cancer and other causes 1897–1997. Br J Radiol 74:507–519
51. Mohan AK, Hauptmann M, Freedman DM et al (2003) Cancer and other causes of mortality among radiologic technologists in the United States. Int J Cancer 103:259–267
52. Miller RW, Jablon S (1970) A search for late radiation effects among men who served as x-ray technologists in the U.S. Army during World War II. Radiology 96:269–274
53. Andersson M, Engholm G, Ennow K et al (1991) Cancer risk among staff at two radiotherapy departments in Denmark. Br J Radiol 64:455–460
54. Yoshinaga S, Aoyama T, Yoshimoto Y, Sugahara T (1999) Cancer mortality among radiological teachnologists in Japan: updated analysis of follow-up data from 1969 to 1993. J Epidemiol 9:61–72
55. Ashmore JP, Krewski D, Zielinski JM et al (1998) First analysis of mortality and occupational radiation exposure based on the National Dose Registry of Canada. Am J Epidemiol 148:564–574
56. Sont WN, Zielinski JM, Ashmore JP et al (2001) First analysis of cancer incidence and occupational radiation exposure based on the National Dose Registry of Canada. Am J Epidemiol 153:309–318
57. Simon SL, Weinstock RM, Doody MM et al (2006) Estimating historical radiation doses to a cohort of U.S. radiologic technologists. Radiat Res 166:174–192
58. Cameron JR (2002) Radiation increased the longevity of British radiologists. Br J Radiol 75:637–640
59. Doody MM, Mandel JS, Lubin JH, Boice JD (1998) Mortality among United States radiologic technologists, 1926–90. Cancer Causes Control 9:67–75
60. Matisane L (2006) Cancer incidence in female health care workers occupationally exposed to ionizing radiation 1982–2002. Health Phys 90(Suppl):S98

# Nuclear Workers

# 6

Most radiation protection agencies deliberately ignore and dismiss radiation hormesis

(Charles Sanders)

Millions of workers employed in the nuclear industry have been exposed to chronic low LET radiation, mostly to cumulative doses <100 mSv [1, 2]. A chronic threshold dose of ~10 mSv/day or ~200 mSv/year was found to cause an excess relative risk (ERR) for all solid cancers in irradiated human populations [3–7]. Luckey was the first to find a biopositive effect of ionizing radiation on cancer formation in nuclear workers (Table 6.1) [1]. A similar study published 17 years later found similar results, that cancer mortality among nuclear workers receiving cumulative lifetime doses of <100 mSv experienced less cancer mortality (Table 6.2).

No relationship was found between radiation exposure and increased cancer incidence in 65 epidemiological studies of populations living around nuclear power stations, fuel reprocessing plants and weapons facilities and testing sites in the U.K., U.S., France and Canada [9]. However, evidence for reduced all cause and all cancer mortality has been found in most epidemiological studies of nuclear workers in scores of locations throughout the world, including nuclear power utility workers, nuclear fuel workers and plutonium workers [10]. The *SMR* for all cancer was 0.74 for combined genders in a cohort of 45,468 Canadian nuclear power industry workers (1957–1994) [11]. A reduced cancer risk (RR = 0.73) was found at cumulative doses of 20–40 mSv in nuclear utility workers of U.S., U.K. and Canada [12, 13].

Up to 175,000 nuclear workers in the UK National Registry for Radiation Workers (NRRW) have been examined in two studies over a 10 year period [14, 15]. Both showed an SMR of about 0.80 [14, 15]. The *SMR* for all cancer at the U.K. Chapelcross nuclear plant was 0.73 [16]. The SMR for all malignant tumors was 0.90 in UKAEA radiation workers compared with nonradiation workers. Substantially lower SMR values were seen for all cause and all cancer during the early years (1946–1953) of employment with the UKAEA when radiation doses were the highest (Fig. 6.1) [17–19].

Mortality was examined in 22,395 nuclear workers at the French National Electricity Company. The mean cumulative dose to the cohort was 19 mSv. A strong beneficial effect was observed that was substantially enhanced with increasing duration of employment (Table 6.3) [20]. Mortality was examined in workers employed in 15 nuclear power utilities in the U.S. between 1979 and 1997. An equally strong beneficial effect was found in

C. L. L. Sanders, *Radiation Hormesis and the Linear-No-Threshold Assumption*,
DOI: 10.1007/978-3-642-03720-7_6, © Springer Verlag Berlin Heidelberg 2010

**Table 6.1** Cancer mortality ratios in male nuclear workers [1]

| Group | Number of workers | | All cancer[a] | Significance |
|---|---|---|---|---|
| | Control | Exposed | | |
| U.S. Shipyard | 31,510 | 38,230 | 0.72 | $p < 0.001$ |
| U.S. Weapons | 20,619 | 15,814 | 0.21 | $p < 0.001$ |
| Canadian Energy | 21,000 | 4,000 | 0.09 | $p < 0.001$ |
| UK Weapons | 24,500 | 70,600 | 0.04 | $p < 0.001$ |

[a]Lowest cancer mortality rate in exposed/controls

**Table 6.2** Cancer mortality ratios in nuclear workers [8]

| Group | Number | PY | Deaths/1,000 | | |
|---|---|---|---|---|---|
| | | | Exposed | Unexposed | Exposed/unexposed |
| Shipyards | 72,356 | 1,591,832 | 9.8 | 13.4 | 0.73 ± 0.04 |
| US DOE | 59,315 | 2,132,046 | 20.8 | 34.8 | 0.60 ± 0.04 |
| Los Alamos | 14,280 | 467,000 | 17.7 | 20.5 | 0.86 ± 0.08 |
| Canada | 8,944 | 268,320 | 20.3 | 23.7 | 0.86 ± 0.11 |
| UK | 95,217 | 3,237,378 | 2.8 | 9.9 | 0.28 ± 0.03 |

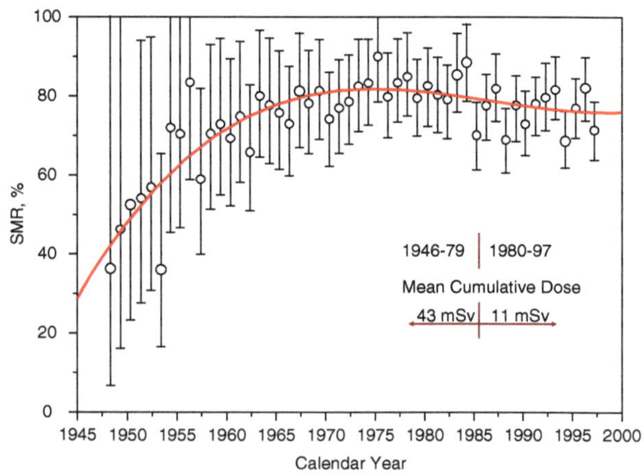


**Fig. 6.1** All cause mortality of employees of the United Kingdom Atomic Energy Authority, 1946–97, compared with the population of England and Wales [17]

53,698 nuclear workers at 15 U.S. nuclear power utilities with a SMR for all solid cancers of 0.65. The mean cumulative dose to the cohort was 26 mSv (Table 6.4) [21].

A study of nuclear power plant workers in Paks, Hungary, showed a strong beneficial effect from radiation exposure. The SMR for all cancer during the period 1985–1998 was 0.68. From 1999 to 2002, the SMR for all causes was only 0.40, while the SMR for all cancer was 0.56 [22]. A large benefit was observed for all cause mortality (SMR = 0.54) and all cancer mortality (SMR = 0.66) in German nuclear power workers; the mean cumulative dose was 31 mSv [23]. The cancer incidence for nuclear workers in Obninsk, Russia

**Table 6.3** SMR values for nuclear workers at the French National Electricity Company; referent was the French general population [20]

| Cause of death | SMR | 95% CI |
|---|---|---|
| All cause | 0.48 | 0.44–0.52 |
| All noncancer | 0.45 | 0.40–0.50 |
| All cancer | 0.58 | 0.49–0.68 |
| Smoking related cancer | 0.50 | 0.39–0.63 |
| Mouth and pharynx | 0.19 | 0.06–0.44 |
| Esophagus | 0.38 | 0.15–0.79 |
| Stomach | 0.41 | 0.11–1.07 |
| Colon | 0.97 | 0.48–1.75 |
| Liver | 0.60 | 0.16–1.54 |
| Nasal cavity | 0.0 | 0.00–0.34 |
| Larynx | 0.76 | 0.38–1.38 |
| Lung | 0.48 | 0.33–0.69 |
| All leukemia | 0.76 | 0.33–1.49 |
| Circulatory diseases | 0.50 | 0.40–0.62 |
| Respiratory diseases | 0.0 | 0.0–0.15 |
| Liver cirrhosis | 0.22 | 0.11–0.39 |

**Table 6.4** SMR values for nuclear workers at 15 U.S. utility sites. The referent group was the U.S. general population [21]

| Cause of death | SMR | 95% CI |
|---|---|---|
| All cause | 0.41 | 0.38, 0.43 |
| All solid cancer | 0.65 | 0.59, 0.72 |
| Stomach cancer | 0.81 | 0.47, 1.32 |
| Colon cancer | 0.75 | 0.53, 1.04 |
| Pancreatic cancer | 0.62 | 0.37, 0.98 |
| Lung cancer | 0.59 | 0.49, 0.71 |
| Prostate cancer | 0.60 | 0.33, 1.01 |
| Kidney cancer | 0.79 | 0.43, 1.32 |
| Central nervous system cancer | 0.85 | 0.54–1.28 |
| All lymphopoietic cancer | 0.65 | 0.48, 0.86 |
| Multiple myeloma | 0.63 | 0.23, 1.37 |
| Leukemia | 1.07 | 0.71, 1.53 |
| All noncancer | 0.34 | 0.32, 0.36 |
| Circulatory system disease | 0.42 | 0.38, 0.47 |
| All respiratory disease | 0.29 | 0.20, 0.40 |
| Digestive system disease | 0.21 | 0.15, 0.30 |

hired before 1981 was compared with the general population of Russia during 1991–1997. The standardized incidence ratio (SIR) in males for all cancers was 0.93 (95% CI: 0.76, 1.12) [24]. A total of 123,661 person-years follow-up was evaluated in Korean nuclear workers. There was no dose–response relationship between cancer incidence and radiation dose. A radiation hormesis response for total cancer incidence was observed at doses <50 mSv (RR = 0.51) and >50 mSv (RR = 0.44) [25]. Cancer mortality of nuclear industry workers in China and Japan showed a radiation hormesis effect [26, 27].

Cancer mortality in nuclear workers at several DOE sites within the U.S. showed either no increase in overall cancer incidence [28] or less than expected cancer incidences [29–33]. The relative risk for all cancers averaged about 0.80. A study of 26,389 employees of Hanford (Richland, WA) who were hired between 1944 and 1978 found evidence of hormesis, particularly for lung cancer mortality [33, 34]. An analysis of cancer mortality was carried out in 106,020 persons employed at Oak Ridge, Tennessee between 1943 and 1984. The *SMR* was 0.80 for all cause mortality and 0.87 for all cancer [35]. The 13 year U.S. Nuclear Shipyard Workers Study (NSWS) evaluated workers health at eight shipyards, finding significantly reduced all cause and all cancer mortality [36, 37].

Cohorts of nuclear workers in 15 countries were evaluated in a pooled data analysis of cancer risk [38, 39]. The study was based on an initial population of about 600,000 nuclear workers. About 32% of all workers were excluded from the analysis. The authors consider the Healthy Worker Survivor Effect (HWSE) as the cause of decreased mortality rates with increasing duration of employment. However, the HWSE is poorly documented and presented as an after-thought with no other alternative explanation other than hormesis (Chap. 7). Cardis et al. claimed a small positive cancer risk [39]. Only ERR values were given as determined by the LNT assumption. Risk by dose level was not given. The authors admit that their data was not well controlled smoking [38]. A significant part of the heterogeneity in epidemiology studies of nuclear workers following low LET exposures is due to smoking and dose-rate effects [40–42].

A companion paper to that of Cardis et al. [39] by Vrijheid et al. [43] presented non-manipulated data on all cause and all cancer mortality for nuclear workers in the same 15 countries. The mean cumulative radiation dose for workers in all 15 countries was 20 mSv [43]. The maximum annual dose was 5.3 mSv. The mortality rates for all age groups were drastically lower than normal. Consistent large decreases in all cause mortality (mean SMR = 0.62) and all cancer mortality (mean SMR = 0.74) were observed (Table 6.5).

The literature was searched and all epidemiological studies that presented an estimate(s) of radiation dose(s) and risk of all cause and all cancer mortality were used in analyses without apparent bias. These studies included populations exposed in the natural environment, during occupational or medical activities, or as a result of accidents. Only non-lagged values were used in the analysis, except where not given (then, 5-year lag data was used). Lagging creates bias in favor of the LNT model by shifting the dose–response curve to the left. It is recommended that no lagging be used, if possible. Confidence Intervals (CI) were mostly at the 95% level, although a few were for 90% limits (all were assumed to be 95%). Studies providing only ERR values using the LNT assumption were ignored.

Radiation doses were evaluated as the group midrange value when given only as a dose range (for example, a midrange of 75 mSv was used for a range of 50–100 mSv). The midrange only indicates the median or average dose in special cases, assuming a known normal distribution. When the highest dose group was expressed in the unbounded form doses > dose x, the dose assignment was subjective, being assigned a value 50% higher than dose x (for example, >100 mSv would be assigned the dose 150 mSv). The risk of all cause mortality and all cancer mortality was expressed as RR. In some cases RR is evaluated indirectly based on Odds Ratio (OR) or SMR. For low probability diseases, RR estimates OR and SMR in a reliable manner. Data and publication sources for Figs. 6.2 through 6.5 are listed in Appendix Tables 6.6 and 6.7. The majority of data is from studies of nuclear workers (Figs. 6.2

**Table 6.5** SMR for all cause and all cancer mortality in nuclear workers employed in 154 facilities in 15 countries. Referent groups were general national populations [43]

| Cohort | Mean dose, mSv | SMR, All cause | 95% CI | SMR, all cancer | 95% CI |
|---|---|---|---|---|---|
| Australia | 6.1 | 0.55 | 0.42–0.71 | 0.65 | 0.41–0.98 |
| Belgium | 26.6 | 0.69 | 0.62–0.77 | 0.62 | 0.50–0.76 |
| Canada | 19.5 | 0.62 | 0.59–0.66 | 0.76 | 0.69–0.84 |
| Finland | 7.9 | 0.86 | 0.77–0.96 | 0.54 | 0.38–0.75 |
| France CEA | 3.8 | 0.59 | 0.55–0.64 | 0.65 | 0.57–0.73 |
| France EDF | 15.8 | 0.49 | 0.44–0.54 | 0.62 | 0.52–0.74 |
| Hungary | 5.1 | 0.40 | 0.33–0.48 | 0.68 | 0.49–0.91 |
| Japan | 18.2 | 0.78 | 0.73–0.82 | 0.87 | 0.79–0.95 |
| South Korea | 15.5 | 0.52 | 0.40–0.67 | 1.03 | 0.65–1.53 |
| Lithuania | 40.7 | 0.40 | 0.33–0.49 | 0.67 | 0.44–0.97 |
| Slovak Republic | 18.8 | 0.53 | 0.37–0.73 | 0.69 | 0.35–1.22 |
| Spain | 25.5 | 0.45 | 0.35–0.57 | 0.57 | 0.38–0.83 |
| Sweden | 17.9 | 0.80 | 0.74–0.86 | 0.95 | 0.82–1.09 |
| Switzerland | 62.3 | 0.77 | 0.60–0.97 | 0.91 | 0.59–1.33 |
| UK | 20.7 | 0.78 | 0.76–0.80 | 0.81 | 0.78–0.84 |
| US Hanford | 23.7 | 0.74 | 0.73–0.76 | 0.80 | 0.76–0.85 |
| US INL | 10.0 | 0.70 | 0.67–0.72 | 0.72 | 0.68–0.77 |
| US NPP | 27.1 | 0.41 | 0/39–0.44 | 0.65 | 0.59–0.73 |
| US ORNL | 15.2 | 0.72 | 0.68–0.77 | 0.82 | 0.72–0.93 |
| MEAN | 20 | 0.62 | – | 0.74 | – |

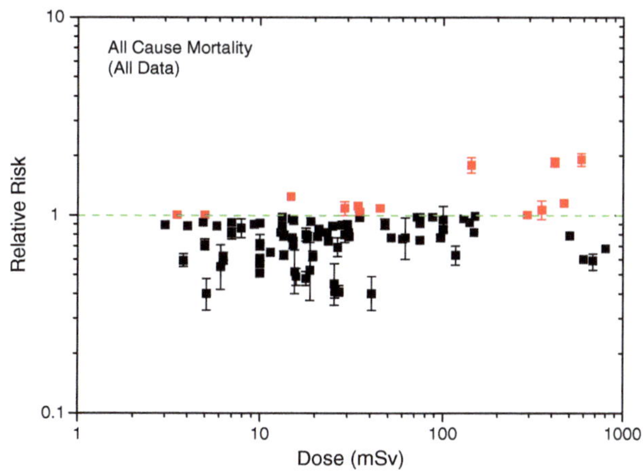


**Fig. 6.2** Relative risk for all cause mortality in all exposed populations

through 6.5). Both nuclear worker and non-nuclear worker populations showed similar beneficial responses. Overwhelming evidence of radiation hormesis was found for all cause mortality and all cancer mortality in both groups of exposed populations at cumulative radiation doses <100 mSv (Figs. 6.2 through 6.5, Appendix Tables 6.6 and 6.7) [44–46].

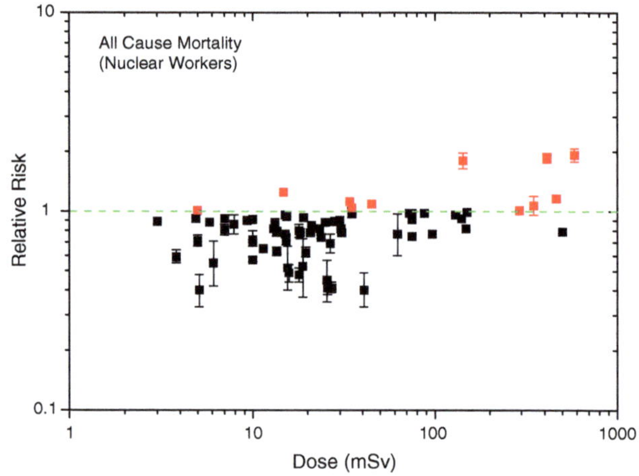


**Fig. 6.3** Relative risk for all cause mortality in nuclear workers

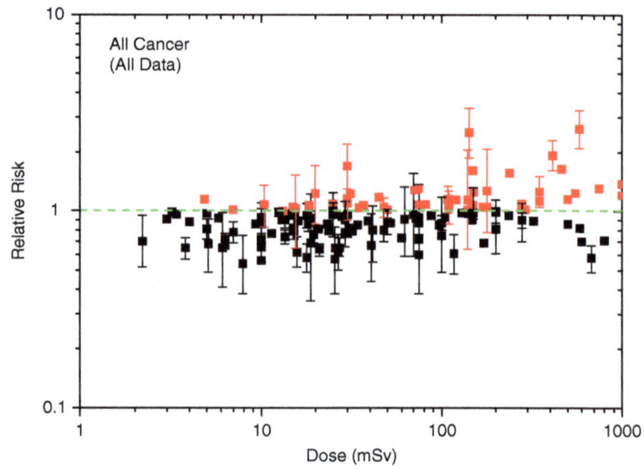


**Fig. 6.4** Relative risk for all cancer mortality in all exposed populations

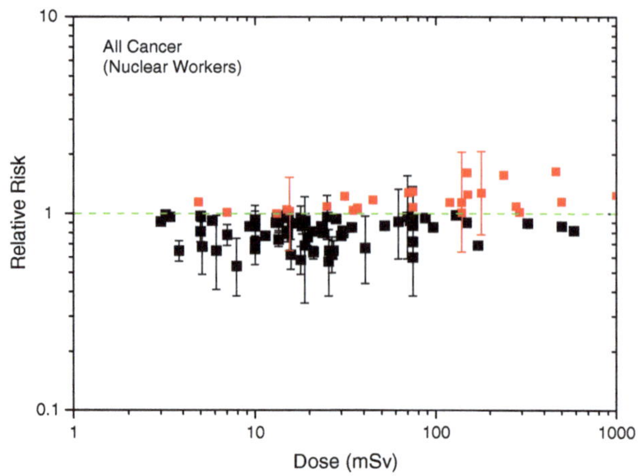


**Fig. 6.5** Relative risk for all cancer mortality in nuclear workers

# Appendix

**Table 6.6** Risk of all cause mortality in epidemiological studies of populations exposed to ionizing radiation

| Dose (mSv) | RR (SMR) | 90–95% CI | Reference |
|---|---|---|---|
| 5 | 0.71 | 0.67–0.76 | Ahrenholz S, Cardarelli J, Dill P et al (2001) Final report. Mortality patterns among workers at the Portsmouth gaseous diffusion plant Piketon, Ohio. NIOSH, Atlanta |
| 13.5 | 0.79 | 0.75–0.83 | Boice JD, Cohen SS, Mumma MT et al (2006) Mortality among radiation workers at Rocketdyne (Atomics International), 1948–1999. Radiat Res 166:98–115 |
| 3 | 0.89 | | |
| 7 | 0.92 | | |
| 25 | 0.88 | | |
| 75 | 0.75 | | |
| 150 | 0.99 | | |
| 100 | 0.85 | 0.82–0.87 | Ivanov VK, Gorski AI, Maksioutov MA et al (2001) Mortality among the Chenobyl emergency workers: estimation of radiation risks (preliminary analysis). Health Phys 81:514–521 |
| 13.5 | 0.63 | 0.60–0.66 | Zablotska LB, Ashmore JP, Howe GR (2004) Analysis of mortality among Canadian nuclear power industry workers after chronic low-dose exposure to ionizing radiation. Radiat Res 161:633–641 |
| 25.7 | 0.41 | 0.38–0.43 | Howe GR, Zablotska LB, Fix JJ et al (2004) Analysis of the mortality experience amongst U.S. nuclear power industry workers after chronic low-dose exposure to ionizing radiation. Radiat Res 162:517–526 |
| 17.9 | 0.48 | 0.44–0.52 | Rogel A, Carre N, Amoros E et al (2005) Mortality of workers exposed to ionizing radiation at the French National Electricity Company. Am J Epidemiol 47:72–82 |
| 678 (lung) | 0.59 | 0.53–0.64 | Wu X, Jiang R (2006) Epidemiology investigate on mortality of uranium miner in Jiangxi Province. Abstracts of the Second Asian and Oceanic Congress for Radiation Protection, Beijing, China, p 354; Wu X (2006) Personal commiunication |
| 13 | 0.82 | – | Schubauer-Berigan MK, Macievic GV, Utterback DF et al (2005) Final report. An epidemiologic study of mortality and radiation-related risk of cancer among workers at the Idaho National Engineering and Environmental Laboratory, a U.S. Department of Energy facility. NIOSH, Atlanta |

*(continued)*

**Table 6.6** (continued)

| Dose (mSv) | RR (SMR) | 90–95% CI | Reference |
|---|---|---|---|
| 3.5 | 1.01 | 0.98–1.05 | Muirhead CR, Bingham D, Haylock RGE et al (2003) Follow up of mortality and incidence of cancer 1952–98 in men from the UK who participated in the UK's atmospheric nuclear weapon tests and experimental programmes. Occup Environ Med 60:165–172 |
| 7 | 0.81 | 0.76–0.86 | Silver SR, Anderson-Mahoney P, Burphy J et al (2005) Mortality update for the Pantex weapons facility: final report health-related energy research branch. National Institute for Occupational Safety and Health, HHS (NIOSH) Publication No. 2005–124 |
| 117 | 0.63 | 0.56–0.70 | Band PR, Le ND, Fang R et al (1996) Cohort study of Air Canada pilots: mortality, cancer incidence, and leukemia risk. Am J Epidemiol 143:137–143 |
| 10,000 | 0.97 | – | Berrington A, Darby SC, Weiss HA et al (2001) 100 years of observation on British radiologists: mortality from cancer and other causes 1897–1997. Br J Radiol 74:507–519 |
| 5,000 | 0.92 | – | |
| 2,000 | 1.00 | – | |
| 800 | 0.68 | – | |
| 10 | 0.91 | – | McGeoghegan D, Binks K (2000) The mortality and cancer morbidity experience of workers at the Capenhurst uranium enrichment facility 1946–95. J Radiol Prot 20:381–401 |
| 140 (lung) | 0.90 | 0.82–0.99 | Veiga LHS, Amaral Didier Colin ECS, Koifman S (2006) A retrospective mortality study of workers exposed to radon in a Brazilian underground coal mine. Radiat Environ Biophys 45:125–134 |
| 19 | 0.93 | 0.89–0.97 | Atkinson WD, Law DV, Bromley KJ et al (2004) Mortality of employees of the United Kingdom Atomic Energy Authority, 1946–97. Occup Environ Med 61:577–585 |
| 5 | 1.01 | – | |
| 15 | 0.96 | – | |
| 35 | 1.04 | – | |
| 75 | 0.98 | – | |
| 130 | 0.96 | – | |
| 97 | 0.77 | 0.74–0.80 | Sponsler R, Cameron JR (2005) Nuclear shipyard worker study (1980–1988): a large cohort exposed to low-dose-rate gamma radiation. Int J Low Radiat 1:463–478 |
| 100 | 0.95 | 0.81–1.11 | Blettner M, Zeeb H, Langner I et al (2002) Mortality from cancer and other causes among airline cabin attendants in Germany, 1960–1997. Am J Epidemiol 156:556–565 |

**Table 6.6** (continued)

| Dose (mSv) | RR (SMR) | 90–95% CI | Reference |
|---|---|---|---|
| 4,000<br>2,500<br>1,300<br>600 | 0.7<br>0.7<br>0.7<br>0.6 | –<br>–<br>–<br>– | Simon SL, Weinstock RM, Doody MM et al (2006) Estimating historical radiation doses to a cohort of U.S. radiologic technologists. Radiat Res 166:174–192<br>Doody MM, Mandel JS, Lubin JH et al (1998) Mortality among United States radiologic technicians, 1926–90. Cancer Causes Control 9:67–75 |
| 45<br>500<br>1,100 | 1.09<br>0.79<br>0.86 | –<br>–<br>– | Wiggs LD, Cox-DeVore CA, Voelz GL (1991) Mortality among a cohort of workers monitored for 210Po exposure: 1944–1972. Health Phys 61:71–76 |
| 21 | 0.63 | 0.60–0.65 | Wiggs LD, Johnson ER, Cox-DeVore CA et al (1994) Mortality through 1990 among white male workers at the Los Alamos National Laboratory: considering exposures to plutonium and external ionizing radiation. 67:577–588 |
| 2,080 (lung) | 0.43 | 0.17–0.88 | Voelz GL, Lawrence JNP, Johnson ER (1997) Fifty years of plutonium exposure to the Manhattan Project plutonium workers: An update. Health Phys 73:611–619 |
| 21 | 0.85 | – | Goldsmith R, Boice JD, Hrubec Z et al (1989) Mortality and career radiation doses for workers at a commercial nuclear power plant: feasibility study. Health Phys 56:139–150 |
| 23.3 | 0.82 | – | Gilbert ES, Omohundro E, Buchanan JA et al (1993) Mortality of workers at the Hanford site: 1945–1986. Health Phys 64:577–590 |
| 5.8 | 0.88 | – | Wilkinson GS, Trieff N, Graham R et al (2000) Final report. Mortality among female nuclear weapons workers. NIOSH, Atlanta |
| 35 | 0.98 | 0.93–1.04 | Carpenter LM, Higgins CD, Douglas AJ et al (1998) Cancer mortality in relation to monitoring for radionuclide exposure in three UK nuclear industry workforces. Br J Cancer 78:1224–1232 |
| 9.3 | 0.90 | – | McGeoghegan D, Binks K (2000) The mortality and cancer morbidity experience of workers at the Springfields uranium production facility, 1946–95. J Radiol Prot 20:111–137 |

*(continued)*

**Table 6.6** (continued)

| Dose (mSv) | RR (SMR) | 90–95% CI | Reference |
|---|---|---|---|
| 4.87 | 0.92 | – | McGeoghegan D, Binks K (2001) The mortality and cancer morbidity experience of employees at the Chapelcross plant of British Nuclear Fuels plc, 1955–95. J Radiol Prot 21:221–250 |
| 14.8 | 1.25 | – | |
| 34.2 | 1.12 | – | |
| 72.0 | 0.98 | – | |
| 148 | 0.82 | – | |
| 292 | 1.01 | – | |
| 464 | 1.16 | – | |
| 60 | 0.76 | | Mohan AK, Hauptmann M, Freedman DM et al (2003) Cancer and other causes of mortality among radiologic technologists in the United States. Int J Cancer 103:259–267 |
| 2,500 | 0.77 | – | Yoshinaga, S, Aoyama T, Yoshimoto Y et al (1999) Cancer mortality among radiological technologists in Japan: updated analysis of follow-up data from 1969 to 1993. J Epidemiol 9:61–72 |
| | | – | |
| 10 | 0.56 | – | Ashmore JP (2000) (unpublished) |
| | | – | |
| 4 | 0.88 | | Caldwell GG, Kelley DB, Zack M et al (1983) Mortality and cancer frequency among military nuclear test participants, 1957 through 1979. JAMA 250:620–624 |
| 10 | 0.64 | 0.56–0.80 | Telle-Lamberton M, Bergot D, Gagneau M et al (2004) Cancer mortality among French Atomic Energy Commission workers. Am J Indust Med 45:34–44 |
| 30 | 0.90 | – | Wiggs LD, Johnson ER, Cox-DeVore CA et al (1994) Mortality through 1990 among white male workers at the Los Alamos National Laboratory: considering exposures to plutonium and external ionizing radiation. Health Phys 67:577–588 |
| 75 | 0.91 | – | |
| 140 | 0.92 | – | |
| 48 | 0.90 | 0.85–0.96 | Dupree-Ellis E, Watkins J, Ingle JN et al (2000) External radiation exposure and mortality in a cohort of uranium processing workers. Am J Epidemiol 152:91–95 |
| 6.3 | 0.61 | 0.57–0.65 | Ashmore JP, Krewski D, Ziellnski JM et al (1998) First analysis of mortality and occupational radiation exposure based on the National Dose Registry of Canada. Am J Epidemiol 148:564–574 |
| 52 | 0.77 | – | Gribbin MA, Weeks JL, Howe GR (1993) Cancer mortality (1956–1985) among male employees of Atomic Energy of Canada Limited with respect to occupational exposure to external low-linear-energy transfer ionizing radiation. Radiat Res 133:375–380 |

**Table 6.6** (continued)

| Dose (mSv) | RR (SMR) | 90–95% CI | Reference |
|---|---|---|---|
| 13.2 | 0.98 | – | Frome EL, Cragle DL, Watkins JP et al (1997) A mortality study of employees of the nuclear industry in Oak Ridge, Tennessee. Radiat Res 148:64–80 |
| 11.4 | 0.65 | – | Artalejo FR, Lara SC, de Andres Manzano B et al (1997) Occupational exposure to ionizing radiation and mortality among workers of the former Spanish Nuclear Energy Board. Occup Environ Med 54:202–208 |
| 15 | 0.77 | – | Beral V, Fraser P, Carpenter L et al (1988) Mortality of employees of the Atomic Weapons establishment, 1951–1982. BMJ 297:757–770 |
| 30.5 | 0.82 | – | Muirhead CR, Goodill AA, Haylock RGE et al (1999) Occupational radiation exposure and mortality:second analysis of the National Registry of Radiation Workers. J Radiol Prot 19:3–26 |
| 87.4 | 0.98 | – | Omar RZ, Barber JA, Smith PG (1999) Cancer mortality and morbidity among plutonium workers at the Sellafield plant of British Nuclear Fuels. Br J Cancer 79:1288–1301 |
| 30.7 | 0.78 | – | Fraser P, Carpenter L, Maconochie N et al (1993) Cancer mortality and morbidity in employees of the United Kingdom Atomic Energy Authority, 1946–1986. Br J Cancer 67:615–624 |
| 15.2 | 0.74 | – | Wing S, Shy CM, Wood JL et al (1991) Mortality among workers at Oak Ridge National Laboratory. Evidence of radiation effects in follow-up through 1984. JAMA 265:1397–1402 |
| 13.2 | 0.88 | – | Loomis DP, Wolf SH (1996) Mortality of workers at a nuclear materials production plant at Oak Ridge, Tennessee, 1947–1990. Am J Ind Med 29:131–141 |
| 28 | 0.89 | – | Rinsky RA, Zumwalde RD, Waxweiler RJ et al (1981) Cancer mortality at a Naval Nuclear Shipyard. Lancet 1:231–235 |
| ?? | 0.62 | – | Wilkinson GS, Tietjen GL, Wiggs LD et al (1987) Mortality among plutonium and other radiation workers at a plutonium weapons facility. Am J Epidemiol 125:231–250 |
| ?? | 0.84 | – | Ritz B (1999) Radiation exposure and cancer mortality in uranium processing workers. Epidemiol 10:531–538 |

*(continued)*

**Table 6.6** (continued)

| Dose (mSv) | RR (SMR) | 90–95% CI | Reference |
|---|---|---|---|
| ?? | 0.82 | – | Hadjimichael OC, Ostfeld AM, D'Atri DA et al (1983) Mortality and cancer incidence experience of employees in a nuclear fuels fabrication plant. J Occup Med 25:48–61 |
| ?? | 0.64 | – | Cragle DL, McLain RW, Qualters JR et al (1988) Mortality among workers at a nuclear fuels production facility. Am J Ind Med 14:379–401 |
| ?? | 1.18 | – | Dupree EA, Cragle DL, McLain RW et al (1987) Mortality among workers at a uranium processing facility, the Linde Air Products Company Ceramics Plant, 1943–1949. Scand J Work Environ Health 13:100–107 |
| 15.3 | 0.94 | 0.90–0.97 | Iwasaki T, Murata M, Ohshima S et al (2003) Second analysis of mortality of nuclear industry workers in Japan, 1986–1997. Radiat Res 159:228–238 |
| 29 | 0.95 | 0.73–1.18 | Zeeb H, Blettner M, Langner I et al (2003) Mortality from cancer and other causes among airline cabin attendants in Europe: a collaborative cohort study in eight countries. Am J Epidemiol 158:35–46 |
| 678 | 0.59 | 0.53–0.64 | Xiao WU, Jiang R, Chang X et al (2006) Epidemiologist investigate on mortality of uranium miner in Jiangxi Province. In: Proceedings of the Second Asian and Oceanic Congress for Radiation Protection, October 9–13, 2006, Beijing, China, pp 1314–1318 |
| 6.1 | 0.55 | 0.42–0.71 | Vrijheid M, Cardis E, Blettner M et al (2007) The 15-country collaborative study of cancer risk among radiation workers in the nuclear industry: design, epidemiological methods and descriptive results. Radiat Res 167:361–379 |
| 26.6 | 0.69 | 0.62–0.77 | |
| 19.5 | 0.62 | 0.59–0.66 | |
| 7.9 | 0.86 | 0.77–0.96 | |
| 3.8 | 0.59 | 0.55–0.64 | |
| 15.8 | 0.49 | 0.44–0.54 | |
| 5.1 | 0.40 | 0.33–0.48 | |
| 18.2 | 0.78 | 0.73–0.82 | |
| 15.5 | 0.52 | 0.40–0.67 | |
| 40.7 | 0.40 | 0.33–0.49 | |
| 18.8 | 0.53 | 0.37–0.73 | |
| 25.5 | 0.45 | 0.35–0.57 | |
| 17.9 | 0.80 | 0.74–0.86 | |
| 62.3 | 0.77 | 0.60–0.97 | |
| 20.7 | 0.78 | 0.76–0.80 | |
| 23.7 | 0.74 | 0.73–0.76 | |
| 10.0 | 0.70 | 0.67–0.72 | |
| 27.1 | 0.41 | 0.39–0.44 | |
| 15.2 | 0.72 | 0.68–0.77 | |

**Table 6.7** Risk of all cancer mortality in epidemiological studies of populations exposed to ionizing radiation

| Dose (mSv) | RR, cancer | 95% CI | Reference |
|---|---|---|---|
| 17.9 | 0.58 | 0.49–0.68 | Rogel A, Carre N, Amoros E et al (2005) Mortality of workers exposed to ionizing radiation at the French National Electricity Company. Am J Epidemiol 47:72–82 |
| 678 (lung) | 0.58 | 0.49–0.67 | Wu X, Jiang R (2006) Epidemiology investigate on mortality of uranium miner in Jiangxi Province. Abstracts of The SecondAsian and Oceanic Congress for Radiation Protection, Beijing, China, p 354; Wu X (2006) Personal commiunication |
| 200 | 0.99 | 0.87–1.14 | Tao Z, Zha Y, Akiba S et al (2000) Cancer mortality in the high background radiation areas of Yangjiang, China during the period between 1979 and 1995. J Radiat Res 41(Suppl):31–41 |
| 20 | 1.23 | 0.86–1.71 | Pukkala E, Auvinen A, Wahlberg G (1995) Incidence of cancer among Finnish airline cabin attendants, 1967–92. Br Med J 311:649–652 |
| 5 | 0.81 | 0.72–0.91 | Ahrenholz S, Cardarelli J, Dill P et al (2001) Final report. Mortality patterns among workers at the Portsmouth gaseous diffusion plant Piketon, Ohio. NIOSH, Atlanta |
| 13.5 | 0.90 | 0.82–0.99 | Boice JD, Cohen SS, Mumma MT et al (2006) Mortality among radiation workers at Rocketdyne (Atomics International), 1948–1999. Radiat Res 166:98–115 |
| 3 | 0.91 | | |
| 7 | 1.02 | | |
| 25 | 1.09 | | |
| 75 | 0.60 | | |
| 150 | 1.61 | | |
| 100 (SIR) | 0.87 | 0.80–0.95 | Ivanov VK, Gorski AI, Maksioutov MA et al (2001) Mortality among the Chenobyl emergency workers: estimation of radiation risks (preliminary analysis). Health Phys 81:514–521 |
| 50 (SIR) | 0.88 | 0.76–1.02 | Ivanov V, Ilyin L, Gorski A et al (2004) Radiation and epidemiological analysis for solid cancer incidence among nuclear workers who participated in recovery operations following the accident at the Chernobyl NPP. J Radiat Res 45:41–44 |
| | 0.7 | 0.52–0.95 | |
| 2.2 | 1.08 | 0.83–1.36 | |
| 10.4 | 0.80 | 0.55–1.06 | |
| 41.5 | 0.98 | 0.91–1.32 | |
| 149.7 all solid tumors | | | |
| 13.5 | 0.74 | 0.68–0.80 | Zablotska LB, Ashmore JP, Howe GR (2004) Analysis of mortality among Canadian nuclear power industry workers after chronic low-dose exposure to ionizing radiation. Radiat Res 161:633–641 |
| 10 | 0.70 | 0.55–0.89 | |
| 70 | 0.96 | 0.59–1.56 | |
| 180 | 1.27 | 0.78–2.07 | |

(*continued*)

**Table 6.7** (continued)

| Dose (mSv) | RR, cancer | 95% CI | Reference |
|---|---|---|---|
| 25.7<br>25<br>75<br>140 | 0.65*<br>0.96<br>0.72<br>1.14 | 0.59–0.72<br>0.74–1.24<br>0.38–1.37<br>0.64–2.05 | Howe GR, Zablotska LB, Fix JJ et al (2004) Analysis of the mortality experience amongst U.S. nuclear power industry workers after chronic low-dose exposure to ionizing radiation. Radiat Res 162:517–526 |
| 109 | 1.15 | 0.98–1.34 | Rahu M, Rahu K, Auvinen A et al (2006) Cancer risk among Chernobyl cleanup workers in Estonia and Latvia, 1986–1998. Int J Cancer 119:162–168 |
| 13 | 0.90 | – | Schubauer-Berigan MK, Macievic GV, Utterback DF et al (2005) Final report. An epidemiologic study of mortality and radiation-related risk of cancer among workers at the Idaho National Engineering and Environmental Laboratory, a U.S. Department of Energy facility. NIOSH, Atlanta |
| 12.5 | 0.99 | 0.94–1.03 | Muirhead CR, Bingham D, Haylock RGE et al (2003) Follow up of mortality and incidence of cancer 1952–98 in men from the UK who participated in the UK's atmospheric nuclear weapon tests and experimental programmes. Occup Environ Med 60:165–172 |
| 7 | 0.78 | 0.69–0.88 | Silver SR, Anderson-Mahoney P, Burphy J et al (2005) Mortality update for the Pantex weapons facility: final report health-related energy research branch. National Institute for Occupational Safety and Health, HHS (NIOSH) Publication No. 2005–124 |
| 117 | 0.61 | 0.48–0.76 | Band PR, Le ND, Fang R et al (1996) Cohort study of Air Canada pilots: mortality, cancer incidence, and leukemia risk. Am J Epidemiol 143:137–143 |
| 10,000<br>5,000<br>2,000<br>800 | 1.75<br>1.24<br>1.12<br>0.71 | –<br>–<br>–<br>– | Berrington A, Darby SC, HA Weiss HA et al (2001) 100 years of observation on British radiologists: mortality from cancer and other causes 1897–1997 Br J Radiol 74:507–519 |
| 140 | 0.97 | – | Ye W, Sobue T, Lee VS et al (1998) Mortality and cancer incidence in Misasa, Japan, a spa area with elevated radon levels. Jpn J Cancer Res 89:789–796 |
| 10 | 0.91 | – | McGeoghegan D, Binks K (2000) The mortality and cancer morbidity experience of workers at the Capenhurst uranium enrichment facility 1946–95. J Radiol Prot 20:381–401 |
| 140 (lung) | 0.73 | 0.55–0.96 | Veiga LHS, Amaral Didier Colin ECS, Koifman S (2006) A retrospective mortality study of workers exposed to radon in a Brazilian underground coal mine. Radiat Environ Biophys 45:125–134 |

**Table 6.7** (continued)

| Dose (mSv) | RR, cancer | 95% CI | Reference |
|---|---|---|---|
| 19 | 0.92 | 0.84–1.00 | Atkinson WD, Law DV, Bromley KJ et al (2004) Mortality of employees of the United Kingdom Atomic Energy Authority, 1946–97. Occup Environ Med 61:577–585 |
| 5 | 0.97 | – | |
| 15 | 0.98 | – | |
| 35 | 1.04 | – | |
| 75 | 1.08 | – | |
| 130 | 0.98 | – | |
| 551 | 1.24 | – | Wang JX, Zhang LA, Li BX et al (2002) Cancer incidence and risk estimation among medical X-ray workers in China, 1950–1995. Health Phys 82:455–466 |
| 82 | 1.08 | | |
| 97 | 0.85 | – | Sponsler R, Cameron JR (2005) Nuclear shipyard worker study (1980–1988): a large cohort exposed to low-dose-rate gamma radiation. Int J Low Radiat 1:463–478 |
| 100 | 0.75 | 0.49–1.18 | Blettner M, Zeeb H, Langner I et al (2002) Mortality from cancer and other causes among airline cabin attendants in Germany, 1960–1997. Am J Epidemiol 156:556–565 |
| 32 | 0.79 (SIR) | 0.77–0.82 | Sont WN, Zielinski JM, Ashmore JP et al (2001) First analysis of cancer incidence and occupational radiation exposurebased on the National Dose Registry of Canada. Am J Epidemiol 153:309–318 |
| 4,000 | 0.9 | – | Simon SL, Weinstock RM, Doody MM et al (2006) Estimating historical radiation doses to a cohort of U.S. radiologic technologists. Radiat Res 166:174–192 |
| 2,500 | 0.8 | – | |
| 1,300 | 0.8 | – | |
| 600 | 0.7 | – | |
| | | | Doody MM, Mandel JS, Lubin JH et al (1998) Mortality among United States radiologic technicians, 1926–90. Cancer Causes Control 9:67–75 |
| 45 | 1.18 | – | Wiggs LD, Cox-DeVore CA, Voelz GL (1991) Mortality among a cohort of workers monitored for 210Po exposure: 1944–1972. Health Phys 61:71–76 |
| 500 | 0.86 | – | |
| 1,100 | 0.66 | – | |
| 21 | 0.64 | 0.59–0.68 | |
| 2,080 (lung) | 0.75 | 0.15–2.18 | Voelz GL, Lawrence JNP, Johnson ER (1997) Fifty years of plutonium exposure to the Manhattan Project plutonium workers: an update. Health Phys 73:611–619 |
| 23.3 | 0.86 | – | Gilbert ES, Omohundro E, Buchanan JA et al (1993) Mortality of workers at the Hanford site: 1945–1986. Health Phys 64:577–590 |
| 500 | 1.15 | – | Shilnikova NS, Preston DL, Ron E et al (2003) Cancer mortality risk among workers at the Mayak Nuclear Complex. Radiat Res 159:787–798 |
| 1,000 | 1.21 | – | |
| 3,000 | 1.85 | – | |
| 5,000 | 1.81 | – | |
| 7,000 | 2.20 | – | |

*(continued)*

**Table 6.7** (continued)

| Dose (mSv) | RR, cancer | 95% CI | Reference |
|---|---|---|---|
| 5.8 | 0.92 | – | Wilkinson GS, Trieff N, Graham R et al (2000) Final Report. Mortality among female nuclear weapons workers. NIOSH, Atlanta |
| 3.4 | 0.96 | – | Carpenter LM, Higgins CD, Douglas AJ et al (1998) Cancer mortality in relation to monitoring for radionuclide exposure in three UK nuclear industry workforces. Br J Cancer 78:1224–1232 |
| 16 | 0.89 | – | |
| 37 | 1.07 | – | |
| 75 | 0.87 | – | |
| 151 | 1.25 | – | |
| 279 | 1.09 | – | |
| 586 | 0.82 | – | |
| 9.3 | 0.86 | – | McGeoghegan D, Binks K (2000) The mortality and cancer morbidity experience of workers at the Springfields uranium production facility, 1946–95. J Radiol Prot 20:111–137 |
| 4.87 | 1.15 | | McGeoghegan D, Binks K (2001) The mortality and cancer morbidity experience of employees at the Chapelcross plant of British Nuclear Fuels plc, 1955–95. J Radiol Prot 21:221–250 |
| 14.8 | 0.95 | | |
| 34.2 | 0.85 | | |
| 72.0 | 0.92 | | |
| 148 | 0.90 | | |
| 292 | 1.02 | | |
| 464 | 1.64 | | |
| 200 | 0.81 (SIR) | 0.61–1.04 | Ron E, Auvinen A, Alfandary E et al (1999) Cancer risk following radiotherapy for infertility or menstrual disorders. Int J Cancer 82:795–798 |
| 25 | 1.00 (SIR) | 1.0–1.1 | Jartti P, Pukkala E, Uitti J et al (2006) Cancer incidence among physicians occupationally exposed to ionizing radiation in Finland. Scand J Work Environ Health 32:368–373 |
| 280 mSv | 0.90 | 0.82–0.98 | Franklyn JA, Maisonneuve P, Sheppard M et al (1999) Cancer incidence and mortality after radioiodine treatment for hyperthyroidism: a population-based cohort study. Lancet 353:2111–2115 |
| 280 mSv | 0.9 | 0.7–1.1 | Goldman MB, Maloof F, Monson RR et al (1988) Radioactive iodine therapy and breast cancer. Am J Epidemiol 127:969–980 |
| 60 | 0.73 (M)<br>0.86 (F) | – | Mohan AK, Hauptmann M, Freedman DM et al (2003) Cancer and other causes of mortality among radiologic technologists in the United States. Int J Cancer 103:259–267 |
| 39 | 0.89 | – | Sun Q et al (2000) Excess relative risk of solid cancer mortality after prolonged exposure to naturally occurring high back-ground radiation of Yangjiang, China. J Radiat Res 41(Suppl):43–52. (All solid cancers only) |
| 105 | 0.93 | – | |
| 171 | 1.05 | – | |
| 237 | 0.95 | – | |

**Table 6.7** (continued)

| Dose (mSv) | RR, cancer | 95% CI | Reference |
|---|---|---|---|
| 50<br>150<br>350<br>750<br>1,500<br>2,500 | 1.02<br>1.07<br>1.13<br>1.31<br>1.74<br>2.16 | | Preston D, Shimizu Y, Pierce DA et al (2003) Studies of the mortality of atomic bomb survivors, Report 13: Solid Cancer and noncancer mortality 1950–97. Radiat Res 160:381–407. (All solid cancers only) |
| 18.4 | 1.07 | – | Andersson M, Engholm G, Ennow K et al (1991) Cancer risk among staff at two radiotherapy departments in Denmark. Br J Radiol 64:455–460 |
| 1,000 | 1.38 | – | Matanoski GM, Sartwell P, Elliott E et al (1984) Cancer risks in radiologists and radiation workers. In: Boice, JD, Fraumeni JF (eds) Radiation carcinogenesis: epidemiology and biological significance. Raven, New York, NY, pp 83–96 |
| 2,500 | 0.81 (M)<br>0.98 (F) | –<br>– | Yoshinaga S, Aoyama T, Yoshimoto Y et al (1999) Cancer mortality among radiological technologists in Japan: updated analysis of follow-up data from 1969 to 1993. J Epidemiol 9:61–72 |
| 6.3 | 0.72 (F)<br>0.68 (M) | 0.66–0.77<br>0.64–0.71 | Ashmore JP, Krewski D, Zielinski JM et al (1998) First analysis of mortality and occupational radiation exposure based on the National Dose Registry of Canada. Am J Epidemiol 148:564–574 |
| 10 | 0.66 (F)<br>0.56 (M) | –<br>– | Ashmore (2000) unpublished |
| 35 | 1.04 | 1.00–1.07 | Sigurdson AJ, Doody MM, Rao RS et al (2003) Cancer incidence in the U.S. radiologic technologists health study, 1983–1998. Cancer 97:3080–3089 |
| 180 | 1.05 | – | Miller RW, Jablon S (1970) A search for late radiation effects among men who served as x-ray technologists in the U.S. Army during World War II. Radiology 96:269–274 |
| 4 | 0.88 Smoky Test | – | Robinette CD, Jablon S, Preston TL (1985) Studies of participants in nuclear tests. National Research Council Follow-Up Agency. |
| 110 | 1.05 | 0.86–1.27 | Reynolds P, Cone J, Layefsky M et al (2002) Cancer incidence in California flight attendants (United States). Cancer Causes Control 13:317–324 |
| ~10 | 0.66 (M)<br>0.93 (F) | 0.63–0.69<br>0.84–1.03 | Telle-Lamberton M, Bergot D, Gagneau M et al (2004) Cancer mortality among French Atomic Energy Commission workers. Am J Indust Med 45:34–44 |

*(continued)*

**Table 6.7** (continued)

| Dose (mSv) | RR, cancer | 95% CI | Reference |
|---|---|---|---|
| 30<br>75<br>140 | 0.77<br>1.30<br>1.01 | –<br>–<br>– | Wiggs LD, Johnson ER, Cox-DeVore CA et al (1994) Mortality through 1990 among white male workers at the Los Alamos National Laboratory: considering exposures to plutonium and external ionizing radiation. Health Phys 67:577–588 |
| 48 | 1.05 | 0.93–1.17 | Dupree-Ellis E, Watkins J, Ingle JN et al (2000) External radiation exposure and mortality in a cohort of uranium processing workers. Am J Epidemiol 152:91–95 |
| 52 | 0.87 | – | Gribbin MA, Weeks JL, Howe GR (1993) Cancer mortality (1956–1985) among male employees of Atomic Energy of Canada Limited with respect to occupational exposure to external low-linear-energy transfer ionizing radiation. Radiat Res 133:375–380 |
| 13.2 | 0.95 | – | Frome EL, Cragle DL, Watkins JP et al (1997) A mortality study of employees of the nuclear industry in Oak Ridge, Tennessee. Radiat Res 148:64–80 |
| 11.4 | 0.77 | – | Artalejo FR, Lara SC, de Andres Manzano B et al (1997) Occupational exposure to ionizing radiation and mortality among workers of the former Spanish Nuclear Energy Board. Occup Environ Med 54:202–208 |
| 15 | 0.82 | – | Beral V, Fraser P, Carpenter L et al (1988). Mortality of employees of the Atomic Weapons establishment, 1951–1982. BMJ 297:757–770 |
| 30.5 | 0.82 | – | Muirhead CR, Goodill AA, Haylock RGE et al (1999) Occupational radiation exposure and mortality:second analysis of the National Registry of Radiation Workers. J Radiol Prot 19:3–26 |
| 3.2<br>14.1<br>31.3<br>71.3<br>120<br>172<br>239<br>324 | .99<br>.97<br>1.23<br>1.28<br>1.14<br>0.69<br>1.57<br>0.89 | –<br>–<br>–<br>–<br>–<br>–<br>–<br>– | Wing S, Richardson DB (2005) Age at exposure to ionizing radiation and cancer mortality among Hanford workers: follow up through 1994. Occup Environ Med 62:465–472 |
| 87.4 | 0.95 | – | Omar RZ, Barber JA, Smith PG (1999) Cancer mortality and morbidity among plutonium workers at the Sellafield plant of British Nuclear Fuels. Br J Cancer 79:1288–1301 |

**Table 6.7** (continued)

| Dose (mSv) | RR, cancer | 95% CI | Reference |
|---|---|---|---|
| 30.7 | 0.80 | – | Fraser P, Carpenter L, Maconochie N et al (1993) Cancer mortality and morbidity in employees of the United Kingdom Atomic Energy Authority, 1946–1986. Br J Cancer 67:615–624 |
| 15.2 | 0.79 | – | Wing S, Shy CM, Wood JL et al (1991) Mortality among workers at Oak Ridge National Laboratory. Evidence of radiation effects in follow-up through 1984. JAMA 265:1397–1402 |
| 13.2 | 1.00 | – | Loomis DP, Wolf SH (1996) Mortality of workers at a nuclear materials production plant at Oak Ridge, Tennessee, 1947–1990. Am J Ind Med 29:131–141 |
| 28 | 0.94 | – | Rinsky RA, Zumwalde RD, Waxweiler RJ et al (1981) Cancer mortality at a Naval Nuclear Shipyard. Lancet 1:231–235 |
| 400 | 0.03 | – | Chen WL, Luan YC, Shieh MC et al (2004) Is chronic radiation an effective prophylaxis against cancer? J Am Physicians Surg 9:6–10 |
| 10 | 0.86 | 0.66–1.10 | Lim YK, Yoo KY (2006) A cohort study on cancer risk by low-dose radiation exposure among radiation workers of nuclear power plants in Korea. J Korea Assoc Radiat Prot 31:53–63 |
| 15.3 | 0.98 | 0.93–1.04 | Iwasaki T, Murata M, Ohshima S et al (2003) Second analysis of mortality of nuclear industry workers in Japan, 1986–1997. Radiat Res 159:228–238 |
| 47.8 | 0.8 | 0.7–1.0 | Hwang SL, Guo HR, Hsieh WA et al (2006) Cancer risks in a population with prolonged low-dose-rate γ-radiation exposure in radiocontaminated buildings, 1983–2022. Int J Radiat Biol 82:849–858 |
| 29 | 0.78 (F)<br>0.90 (M) | 0.66–0.95<br>0.74–1.12 | Zeeb H, Blettner M, Langner I et al (2003) Mortality from cancer and other causes among airline cabin attendants in Europe: A collaborative cohort study in eight countries. Am J Epidemiol 158:35–46 |
| 30 | 1.7 (M)<br>1.1 (F) SIR | 0.9–1.3<br>1.3–2.2 | Haldorsen T, Reitan JB, Tveten U (2001) Cancer incidence among Norwegian airline cabin attendants. Int J Epidemiol 30:825–830 |
| 30 | 0.96 (M)<br>1.16 | –<br>– | Lynge E (2001) Commentary: cancer in the air. Int J Epidemiol 30:830–832 |
| 280 | 1.06 (SIR) | 1.01–1.11 | Hall P, Berg G, Bjelkengren G et al (1992) Cancer mortality after iodine-131 therapy for hyperthroidism. Int J Cancer 50:886–890 |

(*continued*)

**Table 6.7** (continued)

| Dose (mSv) | RR, cancer | 95% CI | Reference |
|---|---|---|---|
| 15 | 1.05 (SIR) | – | Habib RR, Abdallah SM, Law M, Kaldor J (2006) Cancer incidence among Australian nuclear industry workers. J Occup Health 48:358–365 |
| 75 (males) | 0.93 | 0.76–1.12 | Ivanov VK, Tsyb AF, Rastopchin EM et al (2001) Cancer incidence among nuclear workers in Russia based on data from the Institute of Physics and Power Engineering: a preliminary analysis. Radiat Res 155:801–808 |
| 678 | 0.58 | 0.49–0.67 | Xiao WU, Jiang R, Chang X et al (2006) Epidemiologist investigate on mortality of uranium miner in Jiangxi Province. In: Proceedings of the Second Asian and Oceanic Congress for Radiation Protection, Beijing, China, pp 1314–1318 |
| 6.1 | 0.65 | 0.41–0.98 | Vrijheid M, Cardis E, Blettner M et al (2007) The 15-country collaborative study of cancer risk among radiation workers in the nuclear industry: design, epidemiological methods and descriptive results. Radiat Res 167:361–379 |
| 26.6 | 0.62 | 0.50–0.76 | |
| 19.5 | 0.76 | 0.69–0.84 | |
| 7.9 | 0.54 | 0.38–0.75 | |
| 3.8 | 0.65 | 0.57–0.73 | |
| 15.8 | 0.62 | 0.52–0.74 | |
| 5.1 | 0.68 | 0.49–0.91 | |
| 18.2 | 0.87 | 0.79–0.95 | |
| 15.5 | 1.03 | 0.65–1.53 | |
| 40.7 | 0.67 | 0.44–0.97 | |
| 18.8 | 0.69 | 0.35–1.22 | |
| 25.5 | 0.57 | 0.38–0.83 | |
| 17.9 | 0.95 | 0.82–1.09 | |
| 62.3 | 0.91 | 0.59–1.33 | |
| 20.7 | 0.81 | 0.78–0.84 | |
| 23.7 | 0.80 | 0.76–0.85 | |
| 10.0 | 0.72 | 0.68–0.77 | |
| 27.1 | 0.65 | 0.59–0.73 | |
| 15.2 | 0.82 | 0.72–0.93 | |

## References

1. Luckey TD (1991) Radiation hormesis. CRC, Boca Raton, FL
2. Luckey TD (1999) Nurture with ionizing radiation: a provocative assumption. Nutr Cancer 34:1–11
3. Gregoire O, Cleland MR (2006) Novel approach to analyzing the carcinogenic effect of ionizing radiations. Int J Radiat Biol 82:13–19
4. Okamoto K (1987) Critical values of linear energy transfer, dose rates and doses for radiation hormesis. Health Phys 52:671–674

5. Brown SC, Schonbeck MF, McClure D et al (2004) Lung cancer and internal lung doses among plutonium workers at the Rocky Flats plant: a case-control study. Am J Epidemiol 160: 163–172
6. Koshurnikova NA, Bolotnikova MG, Iyin LA et al (1998) Lung cancer risk due to exposure to incorporated plutonium. Health Phys 149:366–371
7. Keirim-Markus IB (2004) Radiation exposure normalization taking account of specific effects at low doses and dose rates. At Energy 93:836–844
8. Luckey TD (2008) Radiation hormesis overview. RSO Mag 8:22–39
9. Shleien B, Ruttenber AJ, Sage M (1991) Epidemiologic studies of cancer in populations near nuclear facilities. Health Phys 61:699–713
10. Sanders CL (2006) Hormesis as a confounding factor in epidemiological studies of radiation carcinogenesis. Korean Assoc Radiat Prot 31:69–89
11. Zablotska LB, Ashmore AP, Howe GR (2004) Analysis of mortality among Canadian nuclear power industry workers after chronic low-dose exposure to ionizing radiation. Radiat Res 161:633–641
12. Cardis E, Gilbert ES, Carpenter L et al (1995) Effects of low doses and low dose rates of external ionizing radiation: cancer mortality among nuclear industry workers in three countries. Radiat Res 142:117–132
13. Carpenter LM, Beral V, Smith PG (1998) Cancer mortality in relation to monitoring for radionuclide exposure in three UK nuclear industry workforces. Br J Cancer 78:1224–1232
14. Muirhead CR, Goodill AA, Haylock RGE et al (1999) Occupational radiation exposure and mortality:second analysis of the National Registry of Radiation Workers. J Radiol Prot 19:3–26
15. Muirhead CR et al (2009) Mortality and cancer incidence following occupational radiation exposure: third analysis of the National Registry for Radiation Workers. Br J Cancer 100:206–212
16. McGeoghegan D, Binks K (2001) The mortality and cancer morbidity experience of employees at the Chapelcross plant of British Nuclear Fuels plc, 1955–95. J Radiol Prot 21:221–250
17. Atkinson WD, Law DV, Bromley KJ et al (2004) Mortality of employees of the United Kingdom Atomic Energy Authority, 1946–97. Occup Environ Med 61:577–585
18. Beral V, Fraser P, Both M et al (1987) Epidemiological studies of workers in the nuclear industry. In: Jones RR, Southwood R (eds) Radiation & health, Wiley, New York, pp 97–106
19. Atkinson WD, Law DV, Bromley KJ et al (2004) Mortality of employees of the United Kingdom Atomic Energy Authority, 1946–97. Occup Environ Med 61:577–585
20. Rogel A, Carre N, Amoros E et al (2005) Mortality of workers exposed to ionizing radiation at the French National Electricity Company. Am J Industr Med 47:72–82
21. Howe GR, Zablotska LB, Fix JJ et al (2004) Analysis of the mortality experience amongst U.S. nuclear power industry workers after chronic low-dose exposure to ionizing radiation. Radiat Res 162:517–526
22. Istvan T, Kerekes A, Otos M, Veress K (2007) A review of cancer mortality data of radiation workers of nuclear power plant, Paks, Hungary, in the light of the International Radiation Epidemiology Study. In: Sixth LOWRAD Conference, Budapest, Hungary, Abstract 122
23. Hammer GP, Fehringer F, Seitz G et al (2008) Exposure and mortality in a cohort of German nuclear power workers. Radiat Environ Biophys 47:95–99
24. Ivanov VK, Tsyb AF, Rastopchin EM et al (2001) Cancer incidence among nuclear workers in russia based on data from the Institute of Physics and Power engineering: a preliminary analysis. Radiat Res 155:801–808
25. Ahn Y, Bae J (2005) A chronic exposure of low-dose radiation and cancer risks among nuclear power plant workers in Korea. In: Proceedings of the 48th Annual Meeting of the Japan Radiation Research Society/the First Asian Congress of Radiation Research, Research Institute for Radiation Biology and Medicine, Hiroshima University, Japan, Abstract S11–S12, 89

26. Sun SQ, Li SY, Yuan LY (1996) Radioepidemiological studies in the nuclear industry of China. Zhonghua Liu Xing Bing Xue Za Zhi 17:333–336
27. Iwasaki T, Murata M, Ohshima S et al (2003) Second analysis of nuclear industry workers in Japan, 1986–1997. Radiat Res 159:228–238
28. Loomis D, Wolf S (1996) Mortality of workers at a nuclear materials production plant in Oak Ridge, Tennessee, 1947–1990. Am J Ind Med 29:131–141
29. Fry SA, Dupree EA, Sipe AH et al (1996) A study of mortality and morbidity among persons occupationally exposed to >50 mSv in a year: Phase I, mortality through 1984. Appl Occup Environ Hyg 11:334–343
30. Wiggs LD, Johnson ER, Cox-DeVore CA et al (1991) Mortality through 1990 among white male workers at Los Alamos National Laboratory: considering exposures to plutonium and external ionizing radiation. Health Phys. 67:577–588
31. Shy C, Wing S (1994) A report on mortality among workers of Oak Ridge National Laboratory: Follow up through 1990. (PO3C-70837, Final Report). Oak Ridge Associated Universities, Oak Ridge, TN, p 21
32. Wilkinson GS, Tietjen GL, Wiggs LD et al (1987) Mortality among plutonium and other radiation workers at a plutonium weapons facility. Am J Epidemiol 125:231–250
33. Gilbert ES, Omohundro E, Buchanan JA et al (1993) Mortality of workers at the Hanford site: 1945–1986. Health Phys 64:577–590
34. Wing S, Richardson D (2005) Age at exposure to ionizing radiation and cancer mortality among Hanford workers: follow-up through 1994. Occup Environ Med 62:465–472
35. Frome EL, DL Cragle DL, Watkins JP et al (1997) A mortality study of employees of the nuclear industry in Oak Ridge, Tennessee. Radiat Res 148:64–80
36. Matanoski GM (1991) Health effects of low-level radiation in shipyard workers. Final Report. Report No. DOE DE-AC02-79EV10095. U.S. Department of Energy, Washington, DC
37. Spousler R, Cameron JR (2005) Nuclear shipyard worker study (1980–1988): a large cohort exposed to low-dose-rate gamma radiation. Int J Low Radiat 1:463–478
38. Cardis E, Vrijheid M, Blettner M et al (2005) Risk of cancer after low doses of ionizing radiation: retrospective cohort study in 15 countries. Br Med J 331:77–80
39. Cardis E, Vrijheid M, Blettner M et al (2007) The 15-country collaborative study of cancer risk among radiation workers in the nuclear industry: estimates of radiation-related cancer risks. Radiat Res 167:396–416
40. Sanders CL, Scott BR (2007) Smoking and hormesis as confounding factors in radiation pulmonary carcinogenesis. Dose Response 6:53–79
41. Thompson RC, Mahaffey JA (1986) Life-span radiation effects studies in animals: whay can they tell us? In: Proceedings of the 22nd Hanford Life Science Symposium, Richland, WA.
42. Stannard JN (1988) Radioactivity and health: a history. Baalman RW (ed) Pacific Northwest Laboratory, National Technical Information Service, Springfield, VA
43. Vrijheid M, Cardis E, Blettner M et al (2007) The 15-country collaborative study of cancer risk among radiation workers in the nuclear industry: design, epidemiological methods and descriptive results. Radiat Res 167:361–379
44. Atkinson WD, Law DV, Bromley KJ (2007) A decline in mortality from prostate cancer in the UK Atomic Energy Authority workforce. J Radiol Prot 27:437–445
45. Wilkinson GS, Trieff N, Graham R et al (2000) Final report. Study of mortality among female nuclear weapons workers. Grant numbers: 1R01 OHO3274, R01/CCR214546, R01/CCR61 2934-01. National Institute for Occupational Safety and Health, Centers for Disease Control and Prevention, Washington
46. Mayya YS (2005) A study of cancer mortality among indian atomic energy. In: Proceedings of the 48th Annual Meeting of the Japan Radiation Research Society/the First Asian Congress of Radiation Research, Research Institute for Radiation Biology and Medicine, Hiroshima University, Japan, Abstract S11–S14, 89–90

# Biased Epidemiological Studies 7

*There are many ways to skin a cat and the cat does not like any of them (Unknown)*

## 7.1 Epidemiology Studies

Observational epidemiological studies are classified as ecological, case–control, or cohort.

In ecologic epidemiological studies, data on populations, rather than data on individuals, are compared. An example of an ecologic study is the evaluation of geographic areas with high-background radiation levels compared with areas with "normal" background levels. Ecological studies aggregate data over a population in a particular area. Ecological studies are subject to problems of correlations between aggregated disease rates and aggregated measures of exposure. Ecological studies compare average exposure with average cancer risk. Advantages of ecological studies are: (1) they are easy and inexpensive; (2) they can document the frequency of disease over time; and (3) they usually include a large population. A good ecological study adequately controls for confounding factors, and has geographic areas with adequate numbers of dose measurement, small variability of dose within individual geographic regions relative to variability in other regions, availability of high-quality health data across geographic regions, and relatively stable populations [7].

Cohort and case–control studies use data for individuals. Case–control studies compare radiation exposure in individuals with cancer and without cancer. In case–control studies, individuals with a specific cancer are compared with a control group of individuals without the cancer with respect to their past exposure to radiation. Case–control studies are usually not used in radiation epidemiology, with the exception for studies of indoor radon and lung cancer. Case–control studies are susceptible to biases of appropriate selection of controls and valid retrospective determination of dose [7].

Cohort studies are prospective in nature and follow cohorts of irradiated populations during their lifetime, relating cumulative exposure with cancer rate. Cohort studies are less

C. L. L. Sanders, *Radiation Hormesis and the Linear-No-Threshold Assumption*,
DOI: 10.1007/978-3-642-03720-7_7, © Springer Verlag Berlin Heidelberg 2010

susceptible to bias from poor recall, take a prolonged period of time, and are more costly. Cohort studies are particularly popular in examination of lung cancer in uranium miners. In retrospective cohort studies, individuals with past radiation exposure are followed, and the rate of cancer is compared between exposed and unexposed individuals or between exposed individuals and the general population [10–12]. Critical elements of a good cohort study are retention of participants through a period of follow-up and good radiation exposure assessment [7].

Proponents of both the LNT assumption and of radiation hormesis have a tendency to "cherry-pick" their epidemiological studies to support their positions. Biased reporting of single and pooled epidemiological studies is a significant problem in clarifying causational relationships in biomedical sciences.

Using LNT assumption causes a strong bias in epidemiological studies [1, 2]. The LNT assumption assumes that cancer risk is proportional to dose. As such, dose (normalized to Sievert) are used as a surrogate for risk. The LNT also assumes that additional doses are additive to risk. Different tissues and organs have different carcinogenic sensitivities to radiation. Different tissue types are assigned tissue weighting factors ($W_I$) that reflect their relative fractional contribution to the total cancer risk. Each $W_I$ is held to be constant, independent of dose, using the LNT assumption. Another assumption is that there is a twofold reduction in cancer risk for a high-dose or high dose-rate exposure, compared with a low-dose or low-dose rate exposure [1].

The literature was searched, and all epidemiological studies that presented an estimate(s) of dose(s) and risk of cancer were used in analyses without apparent bias. These studies included populations exposed in the natural environment, during occupational or medical activities, or as a result of accidents. A simplified, direct approach was used to evaluate the relationship of radiation dose and cancer incidence from ~300 case–control, cohort, and ecologic epidemiological studies in this book, representing many sources and types of exposure (indoor and underground radon, and external low-LET and internal high-LET radiations). Analyses include all cause mortality; all cancer mortality; lung cancer mortality from indoor radon and uranium mining; lung, liver, breast, and CNS cancer mortality; and leukemia mortality following exposure low- and high-LET radiations. Data were given for cumulative radiation dose bins and risk of cancer for each bin in populations exposed to low LET (X-rays and γ-rays) or to high LET (α-particles from Pu, Th, Ra, and Rn) or combinations of both [13, 14].

Radiation doses were evaluated as the group midrange value when given only as a dose range. The midrange only indicates the median or average dose in special cases (e.g., known normal distribution). When the highest dose group was expressed in the unbound form doses >dose X-, the dose assignment was subjective, being assigned a value 50% higher than dose X (e.g., >100 mSv would be assigned the dose 150 mSv). Only nonlagged values were used in the analysis, except where not given (then, 5-year lag data were used [lagging creates bias in favor of the LNT model by shifting the dose–response curve to the left; it is recommend that no lagging be used if possible). Confidence intervals (CI) were mostly at the 95% level, although a few were for 90% limits (all were assumed to be 95%). Where available the CI values were presented in a table and figures. Studies providing only ERR (excess relative risk) values were ignored (see Chapter 6).

## 7.2 Bias, Prejudice, and Statistical Inaccuracy

The design of epidemiological studies for radiation-induced and chemical-induced cancer nearly always uses the LNT assumption. This design makes it difficult to show a threshold or a radiation hormesis response [2, 4, 6]. Using this design often removes evidence of a threshold or of radiation hormesis [9]:

1. Epidemiological researchers somehow came to be convinced that radiation dose was wasted in the carcinogenic process. They felt a need to correct for so-called wasted dose by lagging (throwing away) some of the dose. A 5-year lag means that 5 years of radiation dose for exposed populations is thrown away. Similarly, a 10-year lag means that 10 years of radiation dose is thrown away. Discarding of dose is inconsistent with the LNT assumption. The LNT assumes that each unit of dose is equally capable of causing cancer. Thus, using time lags is entirely inconsistent with the LNT assumption. "This dose lagging trick is still widely used in epidemiological studies but needs to be stopped. Publishers should no longer allow this trick to be used to deceive the readers and funding agencies" [9].

Researchers then calculate the excess relative risk (ERR) per unit dose, which represents the slope of the straight line developed with the LNT assumption. Interestingly, dose lagging is not used when evaluating DNA double-strand break, dose–response curves, since a linear LNT-type dose–response curve is observed at low-radiation doses.

2. Another trick used by epidemiologists is to eliminate the appearance of the hormetic radiation zone by averaging over dose groups, including incorporating low-dose individuals with the control group. Averaging cancer risk appears over a widely varying dose interval that encompasses a threshold and hormesis region, such as 10–150 mSv in Japanese A-bomb survivors. Such dose groups are necessary in case–control studies and often used in cohort studies. Using this trick eliminates the hormetic region in the dose–response curve [9].
3. Constraining the slope of the cancer risk dose–response curve to always be positive is a trick often used in case–control and cohort studies to support the LNT assumption. That often means that low-dose hormetic data that does not fit a positive straight line are simple ignored [9].
4. Ignoring radiation activated natural protection (ANP), which is clearly supported by many low-dose studies, because it does not fit the LNT assumption.

The application of epidemiological models for radiation protection policy is impeded by lack of precision and consistency and individual heterogeneity (genetic sensitivity) to harm or benefit of ionizing radiation. Genetic diversity does not argue against the hormesis concept since sensitive species and individuals at high risk typically display hormetic responses. Hormesis dose–response curves are shifted to the left [2].

Heterogeneity in cancer responses among individuals in populations exposed to ionizing radiation is due to many factors. Among them are variation among tissues and organs, variation in genetic polymorphisms, sensitivity according to age, altered sensitivity due to noncancer health, and confounding factors. Errors in dose estimation are a major source of uncertainty in epidemiological studies of nuclear workers. These errors result from using group means to impute individual doses.

Cancer risk at a small-radiation dose calculated with the LNT assumption is very small. Taking confounding factors, such as smoking, into consideration using appropriate statistical methods does not provide the statistical power in most epidemiological studies to demonstrate causality because factors, because there effects are often much greater than the effects of radiation [20]. Control reference groups should be equally large or larger than the radiation-exposed cohort with a similar distribution of age, gender, and economic and smoking status. The result is that very large numbers of irradiated individuals, at least a million, may be needed to observe statistical significance for increased cancer formation at low doses.

There are a number of statistical, methodological, and very human pitfalls that one encounters while searching for evidence of hormesis in published studies. Epidemiological studies should most closely mimic real-world exposure conditions. Whether hormesis should be the default model, rather than the LNT assumption, in risk assessment and management is the most practical question [1, 4].

For many study designs, it may be more likely for a research claim to be false than true. Ioannidis argued that much of published scientific research is wrong [15, 16]. He studied 49 papers in leading journals that had been cited more than 1,000 times. He found that within a few years, almost one-third of the papers had been refuted by other studies. There is a bias toward publishing positive results, even though negative results are often as informative as positive results. This makes it more likely that incorrect findings end up being published. Preference may be given to studies that show the highest quality of study methods and interpretation, regardless of the results [15, 16].

Conflicts of interest are very common in biomedical research, often prejudiced solely because of a belief in a scientific theory or commitment to the researchers own findings. The "hotter" a scientific field the less likely the research findings are true. A research finding in a biomedical field is less likely to be true: (1) when studies conducted in the field are smaller, (2) when effect sizes are smaller, (3) when there is prejudice, and (4) when more teams are involved in a scientific field in chase of statistical significance. Ubiquitous false-positive claims are more likely in epidemiological studies when the effect size is small. Finally, investigators working in any field are likely to resist accepting that the whole field in which they might have authored or coauthored one-thousand publications and spent their entire career is a "null field" [15, 16, 21].

Claimed research findings may often be simply accurate measures of the prevailing bias. Selective or distorted reporting is a typical form of bias [16]. Bias occurs when experimental design, data analysis, and presentation factors are used that tend to produce research findings that the writer wants. Bias entails manipulation in the analysis and reporting of only part of the findings. Investigators may fail to notice statistically significant relationships or have prejudice that causes them to "bury" significant findings [19].

Controversy is not uncommon among research designs in traditional epidemiological studies [17]. Epidemiological studies with small effects have a high rate of lack of confirmation. The probability that a research finding is true depends on the statistical power of the study and the level of significance. Claims of association with relative risks (RRs) that range from 1.2 to 1.4 may simply be measuring nothing else but the net bias that has been involved in the generation of the data [18].

In epidemiological research, increases in risk of <100% are considered small and difficult to interpret due to chance, statistical bias, and effects of confounding factors. Some journals, such as the *New England Journal of Medicine*, do not accept epidemiological studies for publication unless the RR values are <0.5 or >3.0 [19]. Research findings are more likely to be true where large effects (RR > 3) are observed. An example would be the influence of smoking on lung cancer. Small effect sizes (RR < 3) are likely to be plagued by almost ubiquitous false-positive claims [16].

In addition, prestigious investigators or scientific committees may suppress the peer review process in funding studies and their appearance and dissemination at professional meetings and in publications, condemning their field to perpetuate false dogma [22].

A large number of experimental and epidemiological studies challenge the validity of the LNT assumption, strongly suggesting the presence of a threshold and/or benefits (hormesis) from low-dose, low-LET ionizing radiation [8]. Effects of confounding factors, particularly smoking, and biased reporting of single and pooled epidemiological studies are significant problems in clarifying causational relationships between low-dose radiation and cancer risk.

The advocates of the LNT continue to change strategy, as hormesis researchers continue to establish the facts about the benefits of low-dose radiation. After promoting radiation-induced carcinogenic risks in hundreds of studies using the LNT assumption, they are now taking a different tack. All their studies lack the statistical power to demonstrate harm at doses <100 mSv. Rather than admit to the possibility of a threshold or even benefit at these low doses, they just say they lack the statistical power. The fact is that these studies lack any evidence of harm but abundant statistically significant evidence of benefit at doses <100 mSv.

Many critics deny the reality of hormesis in the published literature. Mechanistic studies with established biological endpoints provide the strongest data that can substantiate or refute the concept of hormesis. For example, responses consistent with hormesis occurred four times more frequently than expected by chance in examining NCI drug-screening databases for 2,189 chemicals [3].

## 7.3 Pooled Studies

Epidemiological data on radiation risks of cancer are highly scattered, making definitive conclusions very difficult. These data, however, are largely consistent with a conclusion that irradiated workers exhibit lower all cause and all cancer mortality than control groups. The large and consistent decreases in mortality cannot be attributable to the HWE.

Efforts have been made to standardize the reporting of meta-analyses to eliminate selective data use and outcome reporting and manipulation of outcomes [17]. Meta-analysis approaches in radiation epidemiology often categorize and trim data or eliminate outlying observations [11]. What categories to use and what data to trim or eliminate provide many opportunities for bias [20, 22]. Meta-analysis findings, with pooling of low-power single studies, often prove to be false. These problems are most often seen in biomedical research

[15, 16, 18]. A meta-analytic finding from inconclusive studies where pooling is used to "correct" the low power of single studies, is probably false; even well-powered epidemiological studies may only have a one in five chance of being true [20].

A meta-analysis of cancer risk using the LNT assumption of over a half-million nuclear workers from 15 countries claimed an ERR for cancers other than leukemia of 0.97 per Sv. Mean cumulative doses ranged from 3.8 to 62.3 mSv. On the basis of this flawed estimate, the authors suggest that 1–2% of deaths from cancer among nuclear workers would be attributable to workplace ionizing radiation [11]. This study is a classic example of biased, unscientific research. The raw data showed a decreased from expected incidence of all cause mortality and all cancer in all nuclear workers from all nations with a mean SMR for all cause mortality of 0.66 and for all cancer mortality of 0.76 [5]. These results were summarily dismissed as the Healthy Worker Effect (HWE) with no documentation and very little discussion (see Chapter 8). A Canadian study was eliminated that, if included, would have made the study null. About 38% of all data was eliminated for various reasons. The reader was not provided with data for all dose bins from each country plotted against cancer incidence. Instead, only a plot of pooled ERR values, which represent the slope of a line for each country obtained by the LNT assumption, was given [11]. An unprejudiced and unabridged examination of all the data, plotting all dose bins with true "zero" dose comparisons using unlagged data, would have clearly demonstrated very significant radiation hormesis.

## References

1. Crump KS (2006) The effect of random error in exposure measurement upon the shape of the exposure response. Dose-Response 3:456–464
2. Calabrese EJ, Baldwin LA (2002) Hormesis and high-risk groups. Reg Toxicol Pharmacol 35:414–428
3. Calabrese EJ, Staudenmayer JW, Stanek EJ, Hoffmann GR (2006) Hormesis outperforms threshold model in NCI anti-tumor drug screening database. Toxicol Sci 94:368–378
4. Elliott KC (2008) Hormesis, ethics, and public policy: an overview. BELLE Newsletter 14:48–50
5. Vrijheid M, Cardis E, Blettner M et al (2007) The 15-country collaborative study of cancer risk among radiation workers in the nuclear industry: design, epidemiological methods and descriptive results. Radiat Res 167:361–379
6. Scott BR, Sanders CL, Mitchel REJ, Boreham DR (2008) CT scans may reduce rather than increase the risk of cancer. J Am Phys Surg 13:7–10
7. Bennett B, Repacholi M, Carr Z (eds) (2006) Health effects of the Chernobyl accident and special health care programmes. World Health Organization, Geneva, p 5
8. Tubiana M (2008) The linear no-threshold relationship and advances in our understanding of carcinogenesis. Int J Low Rad 5:173–204
9. Scott BR (2008) It's time for a new low-dose-radiation risk assessment paradigm—one that acknowledges hormesis. Dose-Response 5:333–351
10. Rosario AS, Wellmann J, Heid IM et al (2006) Radon epidemiology: continuous and categorical trend estimators when the exposure distribution is skewed and outliers may be present. J Toxicol Environ Health A 69:681–700

11. Cardis E, Vrijheid M, Blettner M et al (2007) The 15-country collaborative study of cancer risk among radiation workers in the nuclear industry: estimates of radiation-related cancer risks. Radiat Res 167:396–416
12. Puskin JS (2008) What can epidemiology tell us about risks at low doses? Radiat Res 169:122–124
13. Sanders CL (2006) Hormesis as a confounding factor in epidemiological studies of radiation carcinogenesis. Korean Assoc Radiat Prot 31:69–89
14. Sanders CL, Scott BR (2008) Smoking and hormesis as confounding factors in radiation pulmonary carcinogenesis. Dose-Response 6:53–79
15. Ioannidis JP, Haidich AB, Lau J (2001) Any casualties in the clash of randomized and observational evidence? BMJ 322:879–880
16. Ioannidis JP (2005) Why most published research findings are false. PLOS Medicine 2(8): (http://medicine.plosjournals.org/perlserv?request=get-document&doi=10.1371/journal.pm
17. Vandenbroucke JP (2004) When are observational studies as credible as randomized trials? Lancet 363:1728–1731
18. Wacholder S, Chanock S, Garcia-Closas M et al (2004) Assessing the probability that a positive report is false: An approach for molecular epidemiological studies. J Natl Cancer Inst 96:434–442
19. Topol EJ (2004) Failing the public health-Rofecoxib, Merck, and the FDA. N Engl J Med 351:1707–1709
20. Stroup DF, Berlin JA, Morton SC et al (2000) Meta-analysis of observational studies in epidemiology: a proposal for reporting. Meta-analysis of Observational Studies in Epidemiology (MOOSE) group. JAMA 283:2008–2012
21. Krimsky S, Rothenberg LS, Stott P, Kyle G (1998) Scientific journals and their authors' financial interests: a pilot study. Psychther Psychsom 67:194–201
22. Antman EM, Lau J, Kupelnick B et al (1992) A comparison of results of meta-analyses of randomized control trials and recommendations of clinical experts. Treatments for myocardial infarction. JAMA 268:240–248

# Evidence Negating the Healthy Worker Effect 8

The use of the LNT assumption is "a deeply immoral use of our scientific heritage"

(Lauriston Taylor)

Calabrese in speaking of historical blunders with respect to dose–response relationships in toxicology said: *The effects of this century-long conflict have been as destructive as they have been overlooked, affecting the questions that toxicologists ask and assess, the biological models selected and often the endpoints measured, design of studies, the types of resources needed and employed in toxicological research, exposure standards for carcinogens and non-carcinogens, the cost of environmental and occupational health standards, approaches to risk communication for the general public, and a whole host of clinical opportunities to exploit for patient benefit, amongst others* [1]. Contributing to the dose–response blunders are proponents of the LNT assumption who attribute radiation hormesis to the healthy worker effect (HWE) [2, 3].

The current peer review system for many journals with respect to hormesis is "institutionally" influenced by a type of toxicological political "correctness" in applying the LNT assumption to epidemiological studies of radiation risk [1]. The use of the HWE as a mantra-like explanation for potential benefit from low dose radiation is actually censoring-like behavior that has become routine in many publications without adequate scientific explanation or evidence. Proponents of the LNT consistently consider a positive cancer response as correct and a negative (beneficial) response in need of correction. The HWE is used irrespective of the magnitude of change to avoid invoking the other obvious scientific conclusion, that there is a benefit to be had from exposure to low dose ionizing radiation.

The allegedly stochastic character of cancer induction by ionizing radiation damage to DNA and without the beneficial effects of radiation on cellular defenses, precludes the influence of preemployment health status on HWE. Routine preemployment screening might eliminate a few cancer patients, although this is very unlikely for workers hired before 2000, since genetic prognostic tests were not carried out before then [4]. There were no other medical examinations to discover indicators of future neoplasms. Smoking status was not considered in employment for the vast majority of workers hired before the 1990s, even though smoking status is a very important contributor to cancer risk.

The HWE has been attributed to preemployment medical screening examinations, better working and socioeconomic conditions, annual medical physicals, and superior medical

C. L. L. Sanders, *Radiation Hormesis and the Linear-No-Threshold Assumption*,
DOI: 10.1007/978-3-642-03720-7_8, © Springer Verlag Berlin Heidelberg 2010

care for nuclear workers than in national or other control populations [5–8]. No reduction in mortality from cancer was found in men who received annual medical physicals compared with men who did not [5, 9]. Most nuclear workers from the 1940s to 1980s from which epidemiological studies of cancer risk are derived were men. The two most prevalent cancers in American men are prostate cancer and lung cancer. Numerous randomized control trials of the potential benefit from regular chest X-ray and sputum screening for lung cancer have been carried out. None of them has shown significant benefit compared with nonscreened controls with respect to lung cancer mortality [10–17].

The benefit of screening for prostate cancer with serum prostate-specific-antigen (PSA) and digital rectal examination is unknown [18]. Recently, two large randomized prostate cancer screening trials have been published [19, 20]. A European study found a small reduced mortality rate from prostate cancer but with a high risk of over-diagnosis; a total of 48 additional cases of prostate cancer would have had to be found in 1,410 men of the screened group to prevent one death from prostate cancer [20]. An American study found no difference in mortality rate from prostate cancer. In fact, the incidence of death from prostate cancer was slightly higher in the screening group than in the control group [19]. Thus, it should be obvious that the observed decrease in lung or prostate cancer rates found in nuclear workers was not due to the HWE associated with periodic screening of workers.

The assumption that nuclear workers receive better medical care than those who live and work elsewhere in the same area has not been scientifically demonstrated. Data on how medical screening will somehow remove those who might develop cancer decades later is totally lacking. This does not stop proponents of the LNT and HWE from suggesting that preemployment medical screening somehow identifies future cancer victims [21].

BEIR VII estimates a 15% decrease for all cause mortality as a result of the HWE in nuclear workers exposed to ionizing radiation [22]. Cohort epidemiological studies of occupational nuclear workers commonly exhibit a decrease in all cause mortality rate that is usually much more than 15%, typically 30–50% [23]. The HWE is not observed in case–control studies [22].

The HWE is of no value in interpreting risk of cancer mortality [24]. The HWE is often blamed for lower cancer mortality even when the referent group very closely resembles the radiation exposed group. The HWE is used to explain all types of cancer and all types of diseases. The HWE is presumed not to be applicable when the SMR is 100% or greater, but is presumed to be universally applicable when the SMR is 90, 80 or even 40%. No explanation is given concerning employment medical procedures, as to why the PROFAC is 5% in one country and 60% in another. The level of the HWE does not appear to be important by proponents of the LNT. Thus, there does not appear to be any precise definition of what constitutes the HWE [25].

The concept of the healthy worker survivor effect (HWSE) was adopted in an attempt to explain the decrease in SMRs with increasing duration of employment in nuclear workers throughout the world [23]. The HWSE was proposed to operate when workers in good health stay in employment longer and accumulate higher doses, while those who are sick terminate employment at an earlier date. The HWSE is thought to be due to routine medical screening and dose monitoring for workers employed for long periods [8, 26–28]. Variables such as employment duration, time since hire, age at risk and employment status are all thought to influence the HWSE [26]. Confirming data for this assumption is almost totally lacking [23, 29].

Radiation benefits usually remain unchanged or increase with years of employment in nuclear workers [21, 30]. Epidemiological studies designed to compare exposed and unexposed cohorts in the same company or workplace, where medical procedures for employment and employee health are identical, should best delineate the HWE and HWSE from radiation hormesis [6]. Control groups have been chosen in many studies to eliminate the HWE [6, 31–33].

A very large, pooled, 15-country study of mortality risk in nuclear workers from 154 nuclear facilities, with a mean cumulative dose of 20 mSv was carried out. The consistent, very large decreases in all cause mortality and all cancer mortality of workers in this study (Chap. 6) were claimed to be due to the HWE without providing any confirming evidence [34]. A nonsignificant decrease in all cause mortality was observed with increasing duration of employment (Table 8.1) [23]. An increase in benefit with duration of employment was observed in Pantex nuclear workers [35]. PROFAC values for all cause mortality and all cancer mortality of over 60% cannot be attributed to the HWE or HWSE [36].

The 13 year U.S. Nuclear Shipyard Workers Study (NSWS) evaluated workers health at eight shipyards, finding significantly reduced all cause and all cancer mortality [37, 38]. The study was carried out by Johns Hopkins Department of Epidemiology and a final report

**Table 8.1** PROFAC X 100 (%) for all cause mortality related to duration of employment [23]

| Country cohort | Mean dose (mSv) | Duration of employment | | |
|---|---|---|---|---|
| | | < 5 years | 5–10 years | >10 years |
| Australia | 6.1 | 75 | 57 | 34 |
| Belgium | 26.6 | 32 | 6 | 36 |
| Canada | 19.5 | 37 | 41 | 37 |
| Finland | 7.9 | 0 | 30 | 46 |
| France CEA | 3.8 | 37 | 43 | 41 |
| France EDF | 15.8 | 39 | 53 | 52 |
| Hungary | 5.1 | 78 | 52 | 60 |
| Japan | 18.2 | 8 | 24 | 44 |
| South Korea | 15.5 | 22 | 50 | 75 |
| Lithuania | 40.7 | 61 | 63 | 54 |
| Slovak Republic | 18.8 | 15 | 64 | 50 |
| Spain | 25.5 | 50 | 63 | 52 |
| Sweden | 17.9 | 10 | 22 | 33 |
| Switzerland | 62.3 | 0 | 4 | 49 |
| United Kingdom | 20.7 | 18 | 19 | 25 |
| US Hanford | 23.7 | 27 | 26 | 25 |
| US INL | 10.0 | 32 | 27 | 30 |
| US NPP | 27.1 | 59 | 58 | 59 |
| US ORNL | 15.2 | 31 | 26 | 26 |
| Mean ± std dev | 20.0 ± 13.6 | 32.9 ± 23.2 | 38.3 ± 19.4 | 43.6 ± 13.6 |

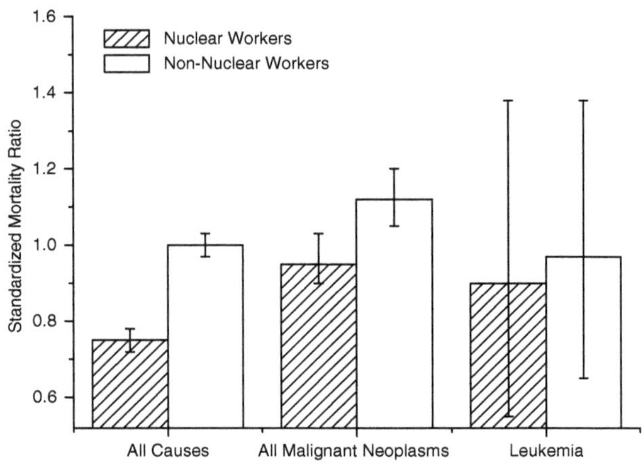


**Fig. 8.1** SMRs for all cause mortality, all malignant tumors and leukemia in U.S. nuclear shipyard workers. The referent was non-nuclear workers working at the same shipyard [38]

written in 1997 [37]. Workers were primarily exposed to $^{60}Co$ gamma rays. The average cumulative shipyard dose was 7.6 mSv/year. The study was designed to avoid selection bias and the HWE in comparing age-matched and job-matched nuclear workers and unexposed controls. A high-dose cohort (>5 mSv) of 27,872, low-dose cohort (<5 mSv) of 10,348 and a control cohort of 32,510 unexposed shipyard workers were examined. The high-dose workers demonstrated significantly lower cancer and all-cause mortality than did unexposed workers. The results showed a statistically significant decrease ($p < 0.001$) for nuclear workers (SMR = 0.76) for all cause mortality as compared with non-nuclear workers (SMR = 1.02) at the same shipyard [38]. For all malignant tumors the *SMR* values were 1.12 (controls), 0.96 (low dose cohort) and 0.95 (high-dose cohort); the *SMR* for the highest dose cohort was significantly less ($p < 0.05$) than controls. The *SMR* for leukemia and hematopoietic cancers was 1.06 (controls), 0.51 (low-dose cohort) and 0.79 (high-dose cohort) (Fig. 8.1) [38].

Wilkinson studied the causes of mortality in women working in twelve U.S. DOE nuclear weapons facilities. The mean radiation dose for workers at all sites was 5.8 mSv [39]. The study covered a total of 67,976 women who worked at these sites before 1980. Wilkinson compared mortality data for female nuclear workers who wore badges to monitor radiation exposure with mortality in female workers who did not wear badges. A strong radiation hormesis response was seen in all facilities for all causes of death and most cancers. Ten of twelve facilities showed decreased lung cancer and eleven of twelve showed decreased breast cancer [39]. There were 25% more deaths from all causes and 17% more deaths from all cancers in unbadged workers than in badged workers. The relative risk for lung cancer mortality in unbadged women who were not monitored was 49% higher than in badged workers (Table 8.2) [39].

All tumor mortality was decreased in monitored nuclear workers employed at Hanford [40] and in monitored nuclear workers at Rocketdyne who had cumulative doses <100 mSv [41]. Both studies used non-monitored workers as referent. A study of nuclear workers at the Idaho National Engineering and Environmental Laboratory (INEEL) compared SMR for all cause mortality in badged workers who received zero dose with badged workers who received a positive occupational dose [42]. All cause mortality and all cancer mortality were significantly less in badged workers with a positive dose than in badged workers with zero dose (Tables 8.3 and 8.4).

**Table 8.2** Ratio of SMRs for cancer in female badged and unbadged nuclear workers. Data is pooled from twelve USDOE facilities (95% CI) [39]

| Cause of death | SMR badged | SMR unbadged | Ratio: badged/ unbadged |
|---|---|---|---|
| All cause | 69 (66–72) | 78 (77–79) | 0.88 |
| All cancer | 71 (66–77) | 77 (74–80) | 0.92 |
| Oral cancer | 61 (24–125) | 103 (78–133) | 0.59 |
| GI cancer | 62 (52–75) | 69 (64–74) | 0.90 |
| Respiratory system cancer | 77 (65–910) | 90 (83–97) | 0.86 |
| Breast cancer | 81 (70–95) | 74 (69–80) | 1.09 |
| Reproductive tract cancer | 58 (45–730) | 72 (66–79) | 0.81 |
| Urinary tract cancer | 79 (47–124) | 82 (66–100) | 0.96 |
| Brain/nervous system cancer | 57 (31–96) | 80 (64–99) | 0.71 |
| Blood/lymph cancer | 73 (56–95) | 79 (70–88) | 0.92 |
| Leukemia | 76 (48–1160) | 83 (70–99) | 0.92 |
| Benign/miscellaneous tumors | 28 (7–730) | 65 (48–85) | 0.43 |

**Table 8.3** Ratio of SMRs for badged workers with positive dose and badged workers with zero dose at INEEL for various causes of death [42]

| Cause of death | SMR badged-zero dose | SMR badged-positive dose | Ratio: positive dose/ zero dose |
|---|---|---|---|
| All Cause | 0.96 | 0.86[a] | 0.90 |
| Diabetes Mellitus | 1.28 | 1.09 | 0.85 |
| Diseases of CNS | 1.32 | 0.92 | 0.70 |
| Diseases Heart | 0.87 | 0.83[a] | 0.95 |
| Disease Circulatory System | 0.98 | 0.81[a] | 0.83 |
| Diseases Respiratory System | 1.05 | 0.81[a] | 0.77 |
| Diseases GI System | 0.95 | 0.69[a] | 0.73 |
| Diseases Genitourinary System | 0.85 | 0.79 | 0.93 |
| Disease Blood-Forming Organs | 0.69 | 0.65 | 0.94 |

[a]Significantly different at $p < 0.05$

SMRs for cancer in 45,468 Canadian nuclear power workers [43] and a combined group of 206,620 Canadian dental, medical, industrial and nuclear power workers [44] gave similar beneficial values. All cause mortality was 0.63 and 0.59 and all cancer mortality was 0.74 and 0.68, respectively. The medically and industrially exposed cohorts showed similar benefits from exposure to radiation as did nuclear workers [43, 44].

UKAEA nuclear workers from 1946 to 1997 were compared with UKAEA non-nuclear workers with respect to mortality. All cause, all cancer and liver, lung, breast and

**Table 8.4** Ratio of SMRs for badged workers with positive dose and badged workers with zero dose at INEEL for cancer deaths [42]

| Cancer deaths | SMR badged-zero dose | SMR badged-positive dose | Ratio: positive dose/zero dose |
|---|---|---|---|
| All cancer | 1.14 | 1.01[a] | 0.89 |
| Buccal cavity | 1.17 | 0.54[a] | 0.46 |
| Digestive tract | 1.20 | 0.99[a] | 0.83 |
| Liver | 1.38 | 1.02 | 0.74 |
| Esophagus | 1.22 | 0.78 | 0.64 |
| Respiratory system | 1.18 | 0.97[a] | 0.82 |
| Larynx | 1.17 | 0.97 | 0.83 |
| Urinary tract | 0.90 | 1.15 | 1.28 |
| CNS | 1.29 | 0.97 | 0.75 |
| Leukemia | 1.00 | 0.98 | 0.98 |

[a]Significantly different at $p > 95\%$

**Table 8.5** Ratio of SMRs for cancer in radiation workers and nonradiation UKAEA workers (95% CI) [45]

| Cause of death | SMR radiation workers | SMR nonradiation workers | Ratio: radiation/nonradiation |
|---|---|---|---|
| All cause | 0.75 (0.74–0.78) | 0.80 (0.78–0.83) | 0.94 |
| All cancer | 0.76 (0.72–0.80) | 0.85 (0.81–0.90) | 0.89 |
| Liver & gallbladder cancer | 0.72 (0.45–1.07) | 0.88 (0.55–1.34) | 0.81 |
| Respiratory system cancer | 0.67 (0.61–0.74) | 0.75 (0.67–0.83) | 0.89 |
| Breast cancer | 0.65 (0.39–1.02) | 0.97 (0.82–1.14) | 0.67 |
| Central nervous system cancer | 0.71 (0.50–0.97) | 0.71 (0.47–1.03) | 1.00 |
| Thyroid cancer | 1.14 (0.31–2.92) | 1.88 (0.75–3.87) | 0.61 |
| Multiple myeloma | 0.65 (0.38–1.04) | 1.19 (0.76–1.77) | 0.55 |
| Leukemia | 1.18 (0.91–1.51) | 1.00 (0.72–1.35) | 1.18 |

thyroid cancer mortality were less in nuclear workers, while leukemia was slightly increased (Table 8.5) [45]. All cause mortality and mortality from smoking-related diseases were decreased at the UK nuclear facilities, Springfields, Capenhurst and Chapelcross, with non-nuclear workers acting as referent (Table 8.6) [46–48]. At all three UK nuclear facilities, the *SMR* for lung cancer in workers with the highest cumulative whole-body dose (400 + mSv) was 0.59 for radiation-monitored workers and 0.97 for unmonitored workers [49]. The authors suggest that the selection process for radiation workers was more stringent than for nonradiation workers [47]. No evidence was given and no explanation offered as to how this would influence cancer incidences decades

**Table 8.6** PROFAC X 100 (%) for radiation workers at UK nuclear facilities involved with nuclear fuel (Springfields), $^{235}U$ diffusion (Capenhurst) and nuclear reactors (Chapelcross). The referent was nonradiation workers at the same sites [46–48]

| Cause of death | Springfields | Capenhurst | Chapelcross |
|---|---|---|---|
| All causes | 14 | 9 | – |
| Disease respiratory system | 21 | 22 | 67 |
| Lung cancer | 3 | 16 | 43 |
| Smoking-related cancers | 8 | 13 | 30 |

later. There is no reason to believe that the HWE is an explanation for these UK studies or for the 20% decline in mortality seen in 175,000 UK National Registry for Radiation Workers [50–52].

British radiologists (1921–1935) were exposed to ~1 Gy/year [53]. The SMR for all cause mortality for those registered with a radiological society after 1920 were very much less than for other medical specialties [54]. U.K. radiologists from 1955 to 1979 accumulated ~50 mSv/year and exhibited an SMR for all cause mortality that was only 0.68 compared with UK nonradiology male physicians; all cancer mortality was 0.71 with no lung cancer deaths in the cohort [54–56]. All medical specialties, like radiologists, have a low mortality ratio from smoking-related diseases [57]. The HWE or smoking habits cannot be seriously considered for this large benefit in all cause and all cancer mortality [50, 58].

Decreased cancer mortality has been observed in radiotherapy patients [59]. All cancer ratios in patients given radioiodine were much less than expected [60, 61]. The reduced cancer ratios below expected rates cannot possibly be explained by the HWE.

A large benefit was seen in French nuclear EDF workers. Mortality was less than half of what was expected, and the effect was greater among workers who spent most of their career in the nuclear sector. The EDF workers and referent experienced the same selection procedures at hiring, medical surveillance procedures, medical care and vacation regime. A longer duration of employment was associated with markedly less all cause and all cancer mortality. SMRs for all cause and all cancer were 0.55 and 0.81, respectively, at <5 year duration of employment, decreasing to 0.47 and 0.57, respectively, at 10+ year duration of employment [62]. The SMR for all cause mortality in German nuclear workers, according to duration of employment, was 0.22 at <5 years, 0.66 at 5–10 years, and 0.54 at >10 years [63]. French nuclear workers had less all cause and all cancer mortality than managers of the same nuclear power facilities. RR for all cause mortality was 0.67 for nuclear workers and 1.19 for managers. All cancer mortality was 0.45 for nuclear workers and 1.05 for managers [64].

Monitored Australian nuclear workers were compared with non-monitored workers in the same plant. The mean dose in exposed workers was 15 mSv. The SIR lung cancer ratio for monitored/non-monitored workers was 0.47 and 0.64 for all smoking-related cancers. Similar results were seen in New South Wales [65]. The *SMR* for cancer for Department of Atomic Energy (DAE) radiation workers in India, aged 20–59 years, was 108 as compared with an *SMR* of 113 for DAE employees who were not radiation workers [66].

Early Soviet nuclear workers at Mayak lived and worked in a dirty, poorly lighted and environmentally toxic chemical and radiation environment, always under the watchful eye of intelligence agents, as a result of Stalin's determined effort to develop nuclear weapons in as short a period of time as possible. A study of all morphologically verifiable lung cancer cases during 1966–1991 in Mayak nuclear workers found a threshold of 0.80 Gy for incorporated $^{239}Pu$; lung and liver tumor incidences were significantly less ($p < 0.05$) than expected levels as was all cause mortality [67, 68]. The claim that emergency workers were subject to additional medical checks before going to work in high radiation zones has even been given as the reason for substantial decreases below the general Russian population in all cause mortality and all cancer mortality (SMRs ~0.80) after the Chernobyl accident [69]. No confirming data was given. The mean stay of the cleanup workers was about 70 days and the mean dose was about 100 mSv [69]. The beneficial results in Chernobyl liquidators and Mayak nuclear workers cannot be attributed to the HWE.

Smoking, alcohol abuse, obesity, high blood pressure, high cholesterol and low consumption of vegetables and fruits are the greatest contributors to disease, while low dose ionizing radiation helps to prevent human disease [70–73]. The beneficial effects of ionizing radiation are remarkably consistent among diverse irradiated human populations exposed as nuclear or medical workers or as inhabitants in high background environments to low dose, low-dose-rate, low-LET radiation exposure [6, 37, 54, 71, 72, 74–76]. Extensive, abundant and diverse human and animal data clearly negates the HWE and HWSE as explanations for radiation hormesis.

## References


1. Calabrese EJ (2007) Historical blunders: how toxicology got the dose-response relationship half right. Cell Mol Biol 51:643–654
2. Howe GR, Chiarelli AM, Lindsay JP (1988) Components and modifiers of the healthy worker effect: evidence from three occupational cohorts and implications for industrial compensation. Am J Epidemiol 128:1364–1375
3. Howe GR, Zablotska LB, Fix JJ et al (2004) Analysis of the mortality experience amongst U.S. nuclear power industry workers after chronic low-dose exposure to ionizing radiation. Radiat Res 162:517–526
4. Kojiro K (1999) The healthy worker effect in a long-term follow-up population. Jpn J Cancer Clin 45:1307–1310
5. Friedman GD, Collen MF, Fireman BH (1986) Multiphasic health checkup evaluation: a 16-year follow-up. J Chronic Dis 39:453–463
6. Luckey TD (1991) Radiation hormesis. CRC, Boca Raton, FL
7. Arrighi HM, Hertz-Picciotto L (1994) The evolving concept of the healthy worker survivor effect. Epidemiology 5:189–196
8. Stewart AM (1990) Healthy worker and healthy survivor effects in relation to the cancer risks of radiation workers. Am J Ind Med 17:151–154
9. Franks P, Gold MR, Clancy CM (1996) Use of care and subsequent mortality: the importance of gender. Health Serv Res 31:347–363
10. Brett GZ (1968) The value of lung cancer detection by six-monthly chest radiographs. Thorax 23:414–420
11. Wilde J (1989) A 10 year follow-up of smi-annual screening for early detection of lung cancer in the Erfurt County, GDR. Eur Respir J 2:656–662

12. Frost JK, Ball WC, Levin ML et al (1984) Early lung cancer detection: results of the initial (prevalence) radiologic and cytologic screening in the Johns Hopkins study. Am Rev Respir Dis 130:549–554
13. Berlin NI (2000) Overview of the NCI Cooperative early lung cancer detection program. Cancer 89:2349–2351
14. Melamed MR (2000) Lung cancer screening results in the National Cancer Institute New York study. Cancer 89:2356–2362
15. Kubik A, Polak J (1986) Lung cancer detection. Results of a randomized prospective study in Czechoslovakia. Cancer 57:2427–2437
16. Manser RL, Irving LB, Byrnes G et al (2003) Screening for lung cancer: a systematic review and meta-analysis of controlled trials. Thorax 58:784–789
17. Bach PB, Kelley MJ, Tate RC, McCrory DC (2003) Screening for lung cancer: a review of the current literature. Chest 123:72S–82S
18. Lin K, Lipsitz R, Miller T, Janakiraman S (2008) Benefits and harms of prostate-specific-antigen screening for prostate cancer: an evidence for update for the U.S. Preventive Services Task Force. Ann Intern Med 149:192–199
19. Schroder FH, Hugosson J, Roobol MJ et al (2009) Screening and prostate-cancer mortality in a randomized European study. N Engl J Med 360:1320–1328
20. Andriole GL, Crawford ED, Grubb RL et al (2009) Mortality results from a randomized prostate-cancer screening trial. N Engl J Med 360:1310–1319
21. Skelcher B (2001) Healthy worker effect. J Radiol Prot 21:71–72
22. Committee to Assess Health Risks from Exposure to Low Levels of Ionizing Radiation (2007) Health risks from exposure to low levels of ionizing radiation: BEIR VII – phase 2. National Academies Press, Washington, DC
23. Vrijheid M, Cardis E, Blettner M et al (2007) The 15-country collaborative study of cancer risk among radiation workers in the nuclear industry: design, epidemiological methods and descriptive results. Radiat Res 167:361–379
24. Li C-Y, Sung F-C (1999) A review of the healthy worker effect in occupational epidemiology. Occup Med 49:225–229
25. Fornalski KW, Dobrzynski L (2010) The healthy worker effect and nuclear industry workers. Dose Response (in press)
26. Arrighi HM, Hertz-Picciotto I (1994) The evolving concept of the healthy worker survivor effect. Epidemiol 5:189–196
27. Steenland K, Deddens J, Salvan A et al (1996) Negative bias in exposure-response trends in occupational studies: modeling the healthy worker survivor effect. Am J Epidemiol 143:202–210
28. Baillargeon J, Wilkinson GS (1999) Characteristics of the healthy survivor effect among male and female Hanford workers. Am J Ind Med 35:343–347
29. Tubiana M, Aurengo A (2006) Dose-effect relationship and estimation of the carcinogenic effects of low doses of ionizing radiation: the Joint Report of the Academiedes Sciences (Paris) and of the Academie Nationale de Medecine. Int J Low Radiat 2:1–19
30. Baillargeon J (2001) Characteristics of the healthy worker effect. Occup Med 16:359–366
31. Luckey TD (2007) Documented optimum and threshold for ionizing radiation. Int J Nuclear Law 1:378–409
32. IDSP (1988) Industrial Disease Standard Panel (ODP). Report No. 3. Report to the workers compensation board on the healthy worker effect, Toronto, Canada
33. UNSCEAR (1994) Annex B: Adaptive responses to radiation in cells and organisms. Report of the United Nations Scientific Committee on the Effects of Atomic Radiation. United Nations, New York, pp 185–272
34. Cardis E, Vrijheid M, Blettner M et al (2007) The 15-country collaborative study of cancer risk among radiation workers in the nuclear industry: estimates of radiation-related cancer risks. Radiat Res 167:396–416

35. Acquavella JF, Wiggs LD, Waxweiler RJ et al (1985) Mortality among workers at the Pantex weapons facility. Health Phys 48:735–746
36. Fornalski KW, Dobrzynski L (2009) Ionising radiation and the health of nuclear industry workers. Int J Low Radiation 6:57–78
37. Matanoski GM (1991) Health effects of low-level radiation in shipyard workers. Final Report. Report No. DOE DE-AC02-79EV10095. U.S. Department of Energy, Washington, DC
38. Spousler R, Cameron JR (2005) Nuclear shipyard worker study (1980–1988): a large cohort exposed to low-dose-rate gamma radiation. Int J Low Radiat 1:463–478
39. Wilkinson GS, Trieff N, Graham R et al (2000) Final Report. Study of mortality among female nuclear weapons workers. Grant Numbers: 1R01 OHO3274, R01/CCR214546, R01/CCR61 2934-01. National Institute for Occupational Safety and Health, Centers for Disease Control and Prevention, Atlanta
40. Gilbert ES, Omohundro E, Buchanan JA et al (1993) Mortality of workers at the Hanford site: 1945–1986. Health Phys 64:577–590
41. Boice JD, Cohen SS, Mumma MT et al (2006) Mortality among radiation workers at Rocketdyne (Atomics International), 1948–1999. Radiat Res 166:98–115
42. Schubauer-Berigan MK, Macievic GV, Utterback DF et al (2005) An epidemiologic study of mortality and radiation-related risk among workers at the Idaho Engineering and Environmental Laboratory, a U.S. Department of energy facility. HHS (NIOSH) Publication No. 2005–131, Cincinnati, OH
43. Zablotska LB, Ashmore JP, Howe GR (2004) Analysis of mortality among Canadian nuclear power industry workers after chronic low-dose exposure to ionizing radiation. Radiat Res 161:633–641
44. Ashmore JP, Krewski D, Ziellnski JM et al (1998) First analysis of mortality and occupational radiation exposure on the National Dose Registry of Canada. Am J Epidemiol 148:564–574
45. Atkinson WD, Law DV, Bromley KJ, Inskip HM (2004) Mortality of employees of the United Kingdom Atomic Energy Authority, 1946–97. Occup Environ Med 61:577–585
46. McGeoghegan D, Binks K (2000) The mortality and cancer morbidity experience of workers at the Springfields uranium production facility, 1946–95. J Radiol Prot 20:111–137
47. McGeoghegan D, Binks K (2000) The mortality and cancer morbidity experience of workers at the Capenhurst uranium enrichment facility 1946–95. J Radiol Prot 20:381–401
48. McGeoghegan D, Binks K (2001) The mortality and cancer morbidity experience of employees at the Chapelcross plant of British Nuclear Fuels plc, 1955–95. J Radiol Prot 21:221–250
49. Carpenter LM, Beral V, Smith PG (1998) Cancer mortality in relation to monitoring for radionuclide exposure in three UK nuclear industry workforces. Br J Cancer 78:1224–1232
50. Simmons JA (2009) Response to 'more on the risk of cancer among nuclear workers'. Letter to the Editor. J Radiol Prot 29:295–296
51. Muirhead CR, Goodill AA, Haylock RGE et al (1999) Occupational radiation exposure and mortality:second analysis of the National Registry of Radiation Workers. J Radiol Prot 19:3–26
52. Muirhead CR et al (2009) Mortality and cancer incidence following occupational radiation exposure: third analysis of the National Registry for Radiation Workers. Br J Cancer 100:206–212
53. Braestrup CB (1958) Past and present radiation exposure to radiologists from the point of view of life expectancy. Am J Roentgen Rad Ther Nucl Med 78:507–519
54. Berrington A, Darby SC, Weiss HA et al (2001) 100 years of observation on British radiologists: mortality from cancer and other causes 1987–1997. Br J Radiol 74:507–519
55. Daunt N (2002) Decreased cancer mortality of British radiologists. Br J Radiol 75:639
56. Smith PG, Doll R (1981) Mortality from cancer and all causes among British radiologists. Br J Radiol 54:187–194
57. Doll R (2005) Mortality of British radiologists: a lecture note. J Radiat Res 46:123–129
58. Daunt N (2002) Decreased cancer mortality of British radiologists. Br J Radiol 75:639–640

59. Luckey TD (2008) Radiation hormesis overview. RSO Mag 8:22–39
60. de Vathaire F, Schumberger M, Delisle MJ et al (1997) Leukaemias and cancers following I-131 administration for thyroid cancer. Br J Cancer 75:734–739
61. Franklyn JA, Maisonneuve P, Sheppard M et al (1999) Cancer incidence and mortality after radioiodine treatment for hyperthyroidism: a population-based cohort study. Lancet 353: 2111–2115
62. Rogel A, Carre N, Amoros E et al (2005) Mortality of workers exposed to ionizing radiation at the French National Electricity Company. Am J Indust Med 47:72–82
63. Hammer GP, Fehringer F et al (2008) Exposure and mortality in a cohort of German nuclear power workers. Radiat Environ Biophys 47:95–99
64. Gros H, Chevalier A et al (2002) Epidemiological surveillance at Electricite de France-Gaz de France: health assessment of nuclear power plant employees between 1993 and 1998. Occup Med 52:35–44
65. Habib RR, Abdallah SM, Law M, Kaldor J (2006) Cancer incidence among Australian nuclear industry workers. J Occup Health 48:358–365
66. Mayya YS (2005) A study of cancer mortality among indian atomic energy. In: Proceedings of the 48th Annual Meeting of the Japan Radiation Research Society/the First Asian Congress of Radiation Research, Research Institute for Radiation Biology and Medicine. Hiroshima University, Japan, Abstract S11–S14; 89–90
67. Tokarskaya ZB, Okladnikova ND, Belyaeva ZD et al (1997) Multifactorial analysis of lung cancer dose-response relationships for workers at the Mayak nuclear enterprise. Health Phys 73:899–905
68. Tokarskaya ZB, Zhuntova GV, Scott BR et al (2006) Influence of alpha and gamma radiations and non-radiation risk factors on the incidence of malignant liver tumors among Mayak workers. Health Phys 91:296–310
69. Ivanov VK, Gorski AI, Maksioutov MA et al (2001) Mortality among the Chernobyl emergency workers: estimation of radiation risks (preliminary analysis). Health Phys 81:514–521
70. Murray CJL, Kulkarni SC, Michaud C et al (2006) Eight Americas: investigating mortality disparities across races, counties, and race-counties in the United States. PloS Med 3(9):e260
71. Sanders CL (2006) Hormesis as a confounding factor in epidemiological studies of radiation carcinogenesis. Korean Assoc Radiat Prot 31:69–89
72. Sanders CL, Scott BR (2008) Smoking and hormesis as confounding factors in radiation pulmonary carcinogenesis. Dose Response 6:53–79
73. Chen WL, Luan YC, Shieh MC et al (2004) Is chronic radiation an effective prophylaxis against cancer? J Am Physicians Surg 9:6–10
74. Cameron JR (2002) Correspondence: radiation increased the longevity of British radiologists. Br J Radiol 75:637–639
75. Mortazavi SMJ, Ikushima T (2006) Open questions regarding implications of radioadaptive response in the estimation of the risks of low-level exposures in nuclear workers. Int J Low Radiat 2:88–96
76. Wilkinson GS, Trieff N, Graham R et al (2000) Final Report. Mortality among female nuclear weapons workers. NIOSH, Atlanta

# 9 Lung Cancer

The most effective method for reducing lung cancer risk from smoking, other than quitting, may be to have an annual whole body CT scan (Charles Sanders)

## 9.1 Introduction

Cancer incidence varies widely among countries of the world with inter- and intra-country incidences of site-specific cancers varying by over 100-fold, correlating with a wide range of dietary and lifestyle variables. Lung cancer accounted for 1.2 million deaths worldwide in 2002, representing 18% of global cancer deaths [1]. Lung cancer is the most prevalent cause of cancer mortality in the U.S., accounting for approximately 30% of all cancer deaths [2, 3]. Exercise, the Mediterranean diet and caloric restriction enhance antioxidant production and immune defenses, decreasing lung cancer risk, while tobacco use increases risk [4,5,6].

## 9.2 Tobacco

Cigarette smoke is a complex mixture of more than 6,000 chemicals that are separated into vapor and particulate phases. Hundreds of chemicals in tobacco smoke react covalently with DNA to form adducts or produce free radicals causing oxidative damage, while ~40 of these compounds are known to cause cancer in humans and/or animals [4]. Cigarette smoke exhibits very significant synergistic interactions with ethanol to induce oral/pharyngeal cancers [7] and with asbestos to induce lung cancer [8]. Environmental exposure to side-stream cigarette smoke may be a significant cause of lung cancer in never smokers [9, 10]. The RR for lung cancer in urban areas of high air pollution is about 1.5 times greater than for rural areas of low air pollution [11].

C. L. L. Sanders, *Radiation Hormesis and the Linear-No-Threshold Assumption*,
DOI: 10.1007/978-3-642-03720-7_9, © Springer Verlag Berlin Heidelberg 2010

Cigarette smoke contains radionuclides. The annual radiation dose to the lung from $^{210}Po$ and $^{210}Pb$ in Italians who smoke 20 cigarettes per day is 0.28 mSv [12]. A heavy smoker may accumulate a lifetime alpha radiation dose of ~1 Gy to bronchial bifurcations, mostly from $^{210}Po$ [5].

The Surgeon General annual report (2004) on smoking and health listed relationships between smoking and cancer in the bladder, cervix, esophagus, kidney, larynx, lung, oral cavity, pharynx, pancreas, and stomach, and for acute myeloid leukemia and Hodgkin's lymphoma [13]. The lung cancer incidence in Dresden, Germany, increased about a 100-fold from 0.06% in 1852 until the 1990s as a result of cigarette production that began in 1862 [14]. About 10% of all lung cancer deaths in the U.S. occur in nonsmokers and about 90% in smokers [15, 16]. About 10% of cigarette smokers will develop lung cancer. Survival from lung cancer is poor with only ~10% of patients living longer than one year after diagnosis.

The risk of lung cancer is dependant on the number of cigarettes smoked per day, use of a filter, type of tobacco, extent of inhalation, number of puffs per cigarette, the length of time smoking, and time since quitting for ex-smokers [17]. The RR for lung cancer in heavy smokers living in the U.S., France, Spain, and Germany ranges from 20–46, and can be >100. Typical RR values for lung cancer mortality associated with ever-smoking, when compared with never-smoking, ranges from 15 to 30 [18]. The *RR* for lung cancer in France was 15.6 for current smokers, 9.2 for ex-smokers who stopped within the previous 10 years, and 2.9 for ex-smokers who stopped at least 10 years earlier; similar incidences of lung cancer are also seen in other western countries [5].

It is difficult to accurately discriminate among different causes of lung cancer at a low tumor prevalence [19]. Useful assessments of radiation risk require accurate estimation of active and passive cigarette smoking exposures and diet [5]. Many epidemiological studies do not provide sufficient smoking stratification cohorts to determine the risk of lung cancer from exposure to ionizing radiation. Variability among epidemiological studies of lung cancer and radiation is mostly due to the powerful confounding influence of cigarette smoking [20–30]. For example, 80% of male Japanese A-bomb survivors reported that they were smokers [31]. Of the about 600 lung cancer cases in Japanese A-bomb survivors, only about 50 were related to radiation exposure [32–34].

## 9.3 External Low LET Radiation

Paracelsus in 1537 was the first to describe a high incidence of lung cancer among miners who were later shown to be exposed to very high levels of radon. However, at lower doses, no one has been identifiably injured by radiation while working within the numerical standards set first by the NCRP and then the ICRP in 1934 [35]. The 1934 ICRP standard was about 500 mSv year$^{-1}$ [36]. The approximate human cumulative threshold dose for lung cancer induction for low-dose-rate, low LET radiation (near continuous) is about 15 Sv [37].

No causal relationship was found between radiation exposure and increased cancer incidence in 65 epidemiological studies of populations living around nuclear power stations, fuel reprocessing plants, and weapons facilities and testing sites in the U.K., U.S.,

France, and Canada [38]. Radiation hormesis for all cancer has been found in a large number of epidemiological studies [4, 5, 39, 40].

Studies of radium dial painters, airline personnel, hard rock and uranium miners, radiologists and radiological technicians, nuclear workers, X-ray technicians, residents with high indoor radon and radium therapy, thorotrast, ankylosing spondylitis patients, Hodgkin disease, pediatric cancer, thymus hypertrophy, tinea capitis, metropathia hemorrhagica, postpartum mastitis, fluoroscopy, peptic ulcer, female breast cancer, and benign breast disease patients have not demonstrated a significant excess of lung cancer at a lung dose <1 Gy in never smokers [5, 41–45]. Average lung doses among Chernobyl liquidators were as high as 0.6 Gy due largely to inhalation of radionuclides [46]; no increase in the incidence of lung cancer was found in any group associated with the Chernobyl accident [47, 48]. A review of the literature led to the conclusion that at a radiation dose of protracted exposure to X- or $\gamma$-rays, <2 Gy, does not cause lung cancer, but often causes a reduction in lung cancer [49].

A significant part of the heterogeneity following low LET exposures is due to cigarette smoking and dose-rate effects [50, 51]. Plotting data from several epidemiological studies, as a daily dose instead of total cumulative dose, removed much of the heterogeneity, with a threshold of >10 mSv day$^{-1}$ for excess relative risk of all solid cancers in irradiated human populations [52]. An earlier study found a dose-rate of 1–10 mGy day$^{-1}$ below which radiation hormesis in the lung was likely to occur [53].

Wilkinson studied the causes of mortality in 67,976 women working at 12 U.S. nuclear weapons facilities before 1980. Mortality data for female nuclear workers who wore badges to monitor radiation exposure was compared with mortality in female workers who did not wear badges. The relative risk of lung cancer mortality in unbadged workers was 49% higher than in badged workers [54]. Both the Shipyard [55] and Wilkinson [54] studies, which had appropriate internal controls demonstrated clear evidence of radiation hormesis for lung cancer.

Nearly 200,000 participants in the National Dose Registry of Canada (1951–1988) were examined for cancer mortality. The PROFAC for lung cancer was 36% in men and 21% in women [56]. In a cohort of 45,468 Canadian nuclear power industry workers (1957–1994), the PROFAC for lung cancer was 19% in males and 60% in females [57]. The PROFAC for lung cancers at the U.K. Chapelcross nuclear plant was 43% ($p < 0.0001$) when compared with Scottish rates and 35% when compared with England and Wales rates [58]. At three U.K. nuclear facilities, the SMR for lung cancer in workers with the highest cumulative whole-body dose (400+ mSv) was 0.59 for monitored workers and 0.97 for unmonitored workers [59]. The PROFAC for lung cancer in UKAEA radiation workers was 11% when compared with non-radiation workers [39].

Mortality was examined in 176,000 Japanese nuclear industry workers from 1986 to 1997. The PROFAC for lung cancer was 57 and 28% at cumulative doses of 50–100 mSv and >100 mSv, respectively [60]. A strong radiation hormesis effect was observed at doses <50 mSv (PROFAC = 49%) and >50 mSv (PROFAC = 56%) for total cancer in Korean nuclear workers [61]. Of the 668 Korean nuclear workers who died during 1984–1999, only 21 died of lung cancer, with a PROFAC for lung cancer of 34% [62]. The PROFAC for lung cancer was 41% in a study of U.S. workers employed at 15 nuclear power plants between 1979 and 1997 [63].

Evidence of radiation hormesis (PROFAC = 13% at 0.01–0.49 Sv, PROFAC = 18% at 0.5–0.99 Sv, and PROFAC = 6% at 1.00–1.99 Sv) was found for lung cancer in fluoroscopy patients being treated for tuberculosis [64, 65]; a similar fluoroscopy study found a PROFAC of 20% at a cumulative lung dose of 0.84 Sv [65] (see Chapter 5). The hormesis effect was similar in men and women. The PROFAC for lung cancer in the contralateral lung at 10 or more years after diagnosis of breast cancer in patients receiving fractionated radiotherapy was 50% at a lung dose of 1.4 Gy [66].

British radiologists employed since 1936 show significantly less than expected incidences of lung cancer, even at annual radiation doses of 70–350 mSv $year^{-1}$ [67]. British radiologists who joined U.K. radiological societies between 1897 and 1997 were divided into four groups depending on when they joined: 1897–1920, 1921–1935, 1936–1954, and 1955–1979. Exposure limits during 1936–1954 were 500 mSv $year^{-1}$, and 50 mSv $year^{-1}$ during 1955–1979. SMR comparisons were made with U.K. male non-radiology physicians. The PROFAC for lung cancer was 26% in the 1936–1954 cohort and 100% for the 1955–1979 cohort (6.5 cases were expected and none observed) [67]. For all cohorts post-1920, the PROFAC for lung cancer was 30% [67]. The PROFAC for lung cancer in U.S. technologists who worked from 1926–1939, 1940–1949, 1950–1959, and 1960–1982 were 28, 24, 17, and 39%, respectively [68]. The PROFAC for lung cancer in Japanese technologists was 38% for those who worked from 1897 to 1933 and 55% for those who worked from 1934 to 1950 [69].

Dose–lung cancer incidence in 54 published epidemiological studies for populations exposed to low LET radiation are presented in Fig. 9.1. The percentage of data points that show RR values for lung cancer <1.0 were 80, 59, and 25% for the 0–100 mSv, 101–1,000 mSv, and >1,000 mSv dose bins, respectively. Although the data were heterogeneous, it showed a pattern of radiation hormesis at doses <1 Sv. For data points where RR < 1.0, the mean RR for the 1–100 mSv dose bin was 0.75 ($n$ = 52) and 0. 67 for the 101–1,000 mSv dose bin ($n$ = 20) (Appendix Table A9.6; Fig. 9.1).

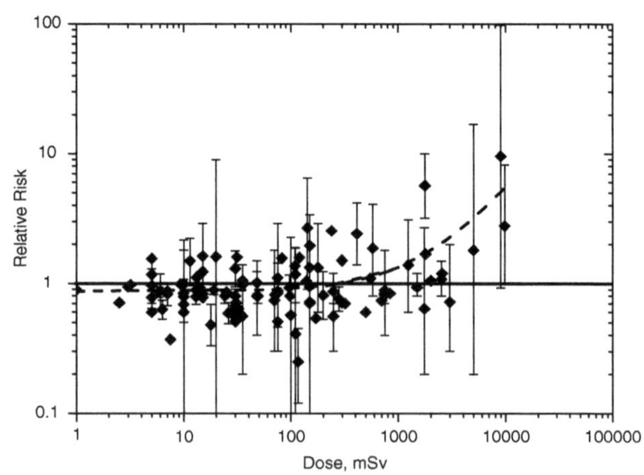


**Fig. 9.1** Lung cancer risk following exposure to low-LET radiation. The equation for the dashed line is Y = 0.87 + 0.000055X; r = 0.64

## 9.4 Radon General

Numerous epidemiological studies of lung cancer risk and indoor radon and radon in underground mines have been published [70–78]. BEIR VI, using the LNT assumption, estimates that 2,100–2,900 never-smokers in the U.S. die each year from radon exposure. The BEIR VI committee does admit that it is especially difficult to estimate radon risks for never smokers in homes. In one meta-analysis, the authors claimed a risk of lung cancer in never smokers while showing no data but a summary line on a figure that was essentially horizontal and zero [79]. Other meta-analyses failed to show any significant association of indoor radon with lung cancer in never smokers [80–82]. The radon database in the IARC (International Agency for Research on Cancer) was reviewed without finding a single study that found radon to be a risk factor for lung cancer in never smokers [82, 83]. The lowest dose for which lung cancer was found in non-smoking U.S. uranium miners was >100 Working Level Months (WLM) [41].

The BEIR VI committee estimated that 10–15% of the annual 160,000 lung cancer deaths in the United States may be attributed to indoor radon, with an uncertainty of 3,300–32,000 deaths, based mostly on an analysis of lung cancer risk in 11 studies of underground miners who received much higher exposures [84]. The mean cumulative exposure among miners was about 30-fold higher than that found in an average home. The BEIR VI committee felt that it is especially difficult to estimate radon risks in homes using high dose data from uranium miners. "Most of the radon-related deaths among smokers would have not occurred if the victims had not smoked." The committee also felt that the assumption of linearity of risk down to the lowest exposures could not be validated against observational data due to the possible presence of a threshold [84].

The Environmental Protection Agency (EPA) action level for residential radon is 4 pCi $L^{-1}$ [49]. The EPA estimates an excess of 29 lung cancer cases in 1,000 persons exposed lifetime to 4 pCi $L^{-1}$ in cigarette smokers and two excess cases in never smokers. In contrast, a recent well-designed case-control study of indoor radon indicated a PROFAC of 62%; accordingly, avoiding indoor radon at 4 pCi $L^{-1}$ would result in 62% excess lung cancers [85, 86].

## 9.5 Environmental and Ecologic Studies of Radon

Normal background radiation exposures are mainly in the range of 2.5–4.0 mSv $year^{-1}$. However, they can exceed 100 times these values in various parts of the world [87]. More than half of the U.S. natural background radiation is associated with exposure to radon (mostly, $^{222}Rn$) and associated daughter radionuclides. Most homes in the U.S. have a radon concentration of about 2 pCi $L^{-1}$ giving a mean dose of 2.2 mSv $year^{-1}$ [88, 89]. The typical lifetime, cumulative, residential radon exposure ranges from 14 to 20 WLM [90].

Radon exposure involves both high-LET α- and low-LET γ-radiations. The γ-ray component is thought to stimulate hormetic effects. The beneficial effects of inhaled and radon-laden water are evident in Russian and European spa hospitals where 100,000s of patients are annually treated for a variety of inflammatory, immune, and hormonal disorders at radon concentrations up to a 1,000 times that of the Environmental Protection Agency (EPA) residential radon concentration limit [14, 91, 92]. Protracted low-dose irradiation with radon enhances cell-mediated immunity and reduces pulmonary metastasis of melanoma in mice [93]. Radon balneology (therapeutic effects of baths) has been shown to be effective in randomized double-blind studies [94]. One study found an optimum therapeutic dose for radon of 2 mSv given over a 2-week period [95]. No increase in lung cancer has been found in radon spa patients or residents living in nearby high background radiation areas.

Relative to other states, Colorado has the third lowest lung cancer death rate in the nation. For the period 1993–1997, the Colorado cancer death rate per 100,000 was 48.2 among males, 25.6 among females. These rates are well below the national averages of 69.4 for males, 34.0 for females per 100,000. Colorado radon levels are well above the national average, averaging 7.3 pCi $L^{-1}$; USEPA estimates the average indoor radon level nationwide is 1.3 pCi $L^{-1}$ [96].

Many reports have shown a negative relationship between environmental radon levels and lung cancer rates (Table 9.1) [97–108]. About 250,000 Americans, living in a handful of Rocky Mountain states where lung cancer rates are much lower than predicted by the U.S. EPA using the LNT assumption, receive high background exposures of ~40 mSv $year^{-1}$ [109] (Table 9.2). Residents near Yangjiang, China, receive an annual background dose of 6.4 mSv. The RR for lung cancer at Yanjiang, China, was 0.81, while the ERR per Sievert was a

**Table 9.1** Correlation of background ionizing radiation levels and lung cancer mortality in 43 U.S. urban populations. No significant association was found for environmental SO4, SO2, NO3, NO2, Cd, Cr, Fe, Ni, or Pb [97]

| Cause of death | Correlation coefficient | Statistical significance |
|---|---|---|
| Lung Cancer | −0.514 | $p < 0.001$ |
| All Cancer | −0.300 | $p < 0.05$ |

**Table 9.2** Comparison of annual cancer mortality (per $10^5$ persons) in Southeastern and Mountain states of the U.S. [109]

| States | All cancer mortality | Lung cancer mortality | Radon level (pCi $L^{-1}$) |
|---|---|---|---|
| Southeastern[a] | 185 | 68 | 0.5 |
| Mountain[b] | 147 | 47 | 2.6 |
| Ratio (mountain/ southeastern) | 0.79 | 0.69 | 5.2 |

[a]Louisiana, Mississippi, Alabama
[b]Idaho, Colorado, New Mexico

**Table 9.3** Lung cancer levels in the Cornwall districts, which experienced the highest radon levels of 55 counties in England and Wales [112]

| Cornwall district | SMR lung cancer | SMR other cancers |
|---|---|---|
| Caradon | 0.73 | 1.01 |
| Carrick | 0.69 | 1.00 |
| Kerrier | 0.80 | 0.95 |
| North Cornwall | 0.66 | 0.92 |
| Penwith | 0.77 | 0.94 |
| Restormel | 0.80 | 1.02 |

negative −0.68 [110]. The incidence of cancer in native populations of Iran, India, and China, exposed to high levels of background radiation, are in most cases decreased [87, 111].

Haynes found a statistically significant negative association between radon concentration and lung cancer in 55 counties of England and Wales (Table 9.3) [112]. Exposure to the highest radon levels in the Cornwall district was associated with the lowest levels of lung cancer with a PROFAC of about 25%. A study in Finland gave similar results [102]. The PROFAC for lung cancer in female non-smokers in four counties of Saxony in East Germany was 40%; the average radon levels exceeded the country average by 3–10 fold [113]. Less than expected lung cancers were found in the NE Hungarian village of Matraderecske at radon levels of 110–165 Bq $m^{-3}$ [22]. In Western Germany, the PROFAC was 19% at >140 Bq $m^{-3}$ with an ERR of −0.02 after adjusting for smoking and asbestos [114]. The relative risk of lung cancer in a radon spa area of Japan was significantly less than in controls with a PROFAC of 45% [115].

Cohen found a powerful protective effect against lung cancer from residential radon exposures [116, 117]. Cohen's study encompassed about 300,000 radon measurements in 1,601 counties of the U.S., representing about 90% of residents living in the U.S. [116]. The trend of county lung cancer mortality in males and females was strikingly negative with increasing radon exposure, even after adjusting for smoking and over 50 other categorizations (Fig. 9.2) [116–121]. For lung cancer in males, the slope was −7.3% per pCi $L^{-1}$ for 1970–1979 data and −7.7% per pCi $L^{-1}$ for 1979–1994 data [120, 122]. Equally strong negative correlations with radon exposure were also found for oral, laryngeal, and

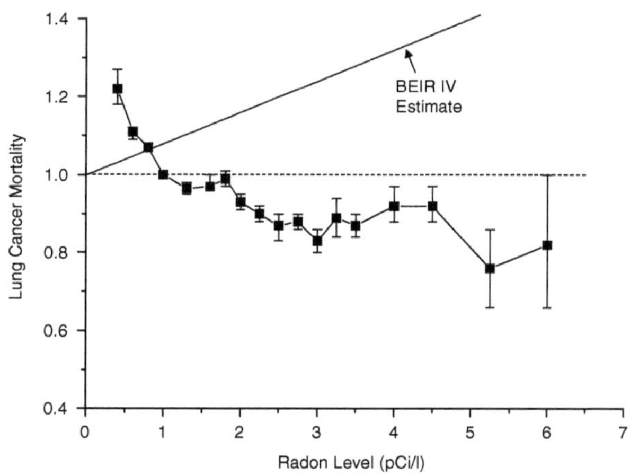


**Fig. 9.2** Lung cancer mortality rates compared with mean home radon levels by U.S. county and compared with the linear model by BEIR IV (RR = ratio of lung cancer mortality rate for residential radon levels to that of average residential level, 1.7 pCi) [116]

**Table 9.4** Estimated change in cancer mortality per unit radon exposure for anatomical sites that exhibit a high smoking-induced cancer mortality [123]

| Cancer site | Gender | Cancers per 100,000 PY |
|---|---|---|
| Lung | M | −8.7 ± 0.4 |
| Lung | F | −9.3 ± 0.5 |
| Oral | M | −10.2 ± 0.7 |
| Oral | F | −11.3 ± 0.9 |
| Larynx | M | −8.5 ± 0.8 |
| Larynx | F | −14.1 ± 1.3 |
| Esophagus | M | −4.1 ± 0.7 |
| Esophagus | F | −7.9 ± 1.1 |

esophageal cancers that are associated with smoking (Table 9.4) [123]. Cohen's data points have very small error bars, which are given in small increments of dose [116]. There was a huge discrepancy of 25 standard deviations between Cohen's results and predictions by BEIR VI using the LNT assumption [124]. At doses >10 pCi $L^{-1}$, the radiation hormesis effect disappears and the lung cancer risk begins to exceed the expected spontaneous incidence. This ecologic study of lung cancer and radon has much greater statistical significance and a sample size that is greater than all other published ecologic studies combined [125]. Cohen provided detailed explanations to critics of his radon findings [124, 126–128].

## 9.6 Case-Control Studies of Radon

Indoor radon, case-control epidemiological studies use different exposure groupings, and different dose bin groupings for exposure groups. The RR is calculated relative to the lowest exposure group, which is different for each study. Most studies do not separate RR values for smokers from nonsmokers [129]. An extrapolation from miner cohort studies may be superimposed on the plotted curve to estimate lung cancer risk. Indoor radon studies are almost all consistent in not finding a dose-dependent association between radon exposure and lung cancer [80]. Case-control studies do not have the statistical power to rule out the presence of a threshold, and are less likely to provide reliable estimates of lung cancer risks from exposure to indoor radon [129]. The results of case-control indoor radon studies are often ambiguous, highly variable, and uncertain [75]. The 95% confidence intervals in some studies are so wide that published plotted ascending lines with an inconsistent dose–response are actually more compatible with a descending line [130].

BEIR VI has inappropriately extrapolated radon-related lung cancer risk from high dose, high dose-rate occupational (underground mines) exposures to low dose, low dose-rate indoor radon exposures [131–133]. Estimates of excess lung cancer risk using the LNT assumption at exposures less than threshold values, such as ~800 WLM threshold in German miners [134], are not credible. The LNT assumption clearly does not best represent the true dose–response relationship of epidemiological data in either individual or pooled studies of indoor radon. Data points from most case-control indoor radon and cohort miner studies are incapable of demonstrating nil, harmful, or beneficial effects at exposures <100 WLM [75, 76, 135]. There is little scientific, statistically valid data that a dose–response or threshold does or does not exist for indoor radon [135, 136]. Several case-control epidemiological studies have shown a negative ERR [71, 114, 130, 137–139].

The combined effect of indoor radon and smoking has not been clarified. The ability of indoor radon to cause lung cancer in never smokers has not been demonstrated [79, 82, 83, 140, 141]. No significant risk of lung cancer was found in never smokers in two large meta-analysis studies of indoor radon [76, 79]. Detailed data on smoking status and dose groupings was not given in a number of studies [82].

Twenty indoor radon studies up to 2007 were evaluated. The lowest and highest lung cancer risks appeared randomly distributed among the exposure cohorts constituting each study. In only 7 (35%) was the lowest exposure ($Bq\,m^{-3}$) cohort associated with the lowest risk of lung cancer. Likewise, in only 10 (50%) studies was the highest exposure cohort associated with the highest risk. Data from 70 dose points were evaluated in the 20 studies of indoor radon. Of these, 20/70 (28.6%) exhibited an RR of ≤ 1.0, while only 6/70 (8.6%) had an RR of >2.0 and <3.0; three of these were in one study [142]. The lowest RR value was 0.25 [115] and the highest was 2.96 [142]. Only three studies showed a consistent, but weak, dose–response; two were positive [143, 144] and one was negative [145]. The 95% CI were large in all studies. A near straight horizontal line best fit the data for indoor radon ($Y = 1.05 + 0.000058X$ with $R = 0.028$) in these 20 studies. Overall, there was no relationship between radon exposure and RR of lung cancer, either benefit or harm (Appendix Table A9.7; Fig. 9.3).

Pooling several single studies in meta-analyses has not been successful in removing heterogeneity of indoor radon data [72, 76, 77, 79]. Pooled analyses from indoor radon studies are unable to rule out the existence of a threshold [146]. Meta-analyses of lung cancer risk and radon exposure often fail to provide relative risk data for dose-exposure groups in each study, giving only the excess odds ratio at $100\,Bq\,m^{-3}$ obtained by regression using the LNT assumption [72, 79, 140]. Meta-analyses of indoor radon and lung cancer often show significant differences between the studies in exposure–response relationships, including evidence

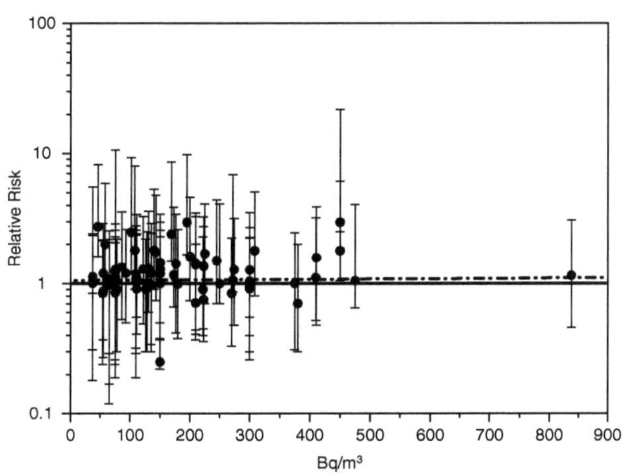


**Fig. 9.3** Case–control studies of the relationship between indoor radon concentration and relative risk of lung cancer. The equation for the fitted dashed straight line was $Y = 1.05 + 0.0002X$. The correlation coefficient was 0.028

of hormesis [76, 81, 141]. Nine of the 21 listed case-control studies of indoor radon through 2007 showed at least one dose bin with non-significant decreases in lung cancer risk.

A meta-analysis of eight case-control indoor radon studies barely excluded "no effect" [80]. A meta-analysis of 17 case-control studies found that "no definitive conclusions could be drawn on the role of radon residential exposure on the risk of lung cancer" [77]. A meta-analysis of 13 European case-control studies showed high heterogeneity; 2 of 13 studies exhibited a decrease in risk, 10 of 13 studies had a minimum 95% CI that was <1.0, while the risk per 100 Bq $m^{-3}$ for 11 studies that showed a positive risk ranged over a factor of >10 from lowest to highest estimate [79]. No individual dose bin data were given for each reported study in several meta-analysis studies [72, 76, 79]. A meta-analysis of 21 case-control studies of indoor radon found negative ERR values for lung cancer for 1950–1954, 1978–1987, and 1988–1997 cohorts; positive results were found in the other cohorts of the 1950–1997 time frame [147].

A very well-designed (controlled for nine categories of smoking and continuous radon monitoring) case-control study of lung cancer risk from indoor radon showed a significant reduction in risk (Table 9.5; Fig. 9.4). AORs (Adjusted Odds Ratio) for lung cancer at 25, 50, 75, 150, and 250 Bq $m^{-3}$ were 0.53, 0.31, 0.47, 0.22, and 2.50, respectively [90]. According to the authors, *this result was entirely unexpected.* Lifetime exposure to residential radon at the EPA's action level of 4 pCi $L^{-1}$ was associated with a ~60% reduction in lung cancer in this study [86].

**Table 9.5** A very well-designed case–control study of lung cancer, adjusted for nine categories of smoking status and year-long radon dosimetry [90]

| Radon (Bq $m^{-3}$) | Adjusted odds ratio | 95% CI |
|---|---|---|
| <25 | 1.00 | – |
| 25–49 | 0.53[a] | 0.24–1.13 |
| 50–74 | 0.31[b] | 0.13–0.73 |
| 75–149 | 0.47[a] | 0.20–1.10 |
| 150–249 | 0.22[a] | 0.04–1.13 |
| >249 | 2.50 | 0.47–13.46 |

[a] $p < 0.1$
[b] $p < 0.05$

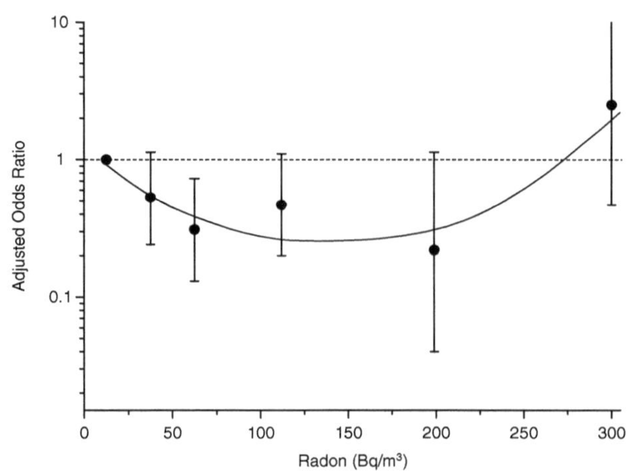


**Fig. 9.4** Case–control study of lung cancer risk from residential radon exposure in Worcester County, Massachusetts; data are for adjusted odds ratio with 95% CI [90]

## 9.7 Underground Uranium Miners

Exposure to radon in underground mines is associated with an increased risk of lung cancer [84]. The proportion of uranium miners who are smokers is particularly high [134, 148]. However, good smoking data are lacking in most studies of uranium miners [23, 149]. Exposure to uranium ore dust and γ-rays may contribute 25–75% of the "effective" dose to the lung in uranium mines. Applying this observation decreases the risk estimates of lung cancer from radon by a factor of 2–3, increasing the likelihood of thresholds [150]. Smoking-related small cell carcinoma represents the majority of lung cancer cases in American uranium miners [84, 151]. Most cohort studies of uranium miners use the LNT assumption to estimate lung cancer risk [75, 80, 84].

Other potential carcinogens, in addition to radiation from radon daughters in uranium mines, are uranium ore dust, silica, and diesel exhaust [152]. Inhaled crystalline silica is a human carcinogen [153]. A lung cancer SMR of 3.1 was associated with an SMR of 39 for silicosis in one study of Newfoundland Fluorspar miners who received a mean cumulative radon exposure of 378 WLM [23].

Evaluation of 11 underground miner studies showed a highly variable, dose–response relationship between radon dose (WLM) and RR of lung cancer. Values for ERR/100 WLM for lung cancer ranged 32-fold from lowest to highest [84]. The RR for lung cancer was ~25–29 in uranium miners with cumulative exposures of >1,450 WLM when compared with those exposed to <80 WLM [75, 154]. The lowest dose for which lung cancer was found in non-smoking U.S. uranium miners was ~200 WLM [154]. The lowest dose of lung cancer seen in one group of non-smoking uranium miners was 465 WLM [155]. The threshold for lung cancer in all U.S. uranium miners was >80 WLM [75, 154] >40 WLM in Australian uranium miners [156], >60 WLM in French and Czech uranium miners [157], >600 WLM in Chinese tin miners [158], and >800 WLM in German uranium miners [134] (Figs. 9.5 and 9.6). The lung seems to be the only organ at risk for cancer from radon in underground mines, although a few studies showed evidence of hormesis with RR values for all non-lung cancers ranging from 0.73 to 0.89 [150, 159].

## 9.8 Internal High LET Radiation

Substantial α-dose thresholds have been found in human populations; for bone tumors at 10 Gy to skeleton from radium [160] and for liver tumors at 2 Gy to liver from thorotrast [161, 162] and plutonium [163]. Similar observations have been made for lung tumors following inhalation of α-emitters.

A cohort of 4,402 workers at the Mound Facility were chronically exposed to $^{210}$Po. The PROFACs for lung cancer at the two highest dose groups were 46 and 66% [164]. U.K. workers at the Springfields uranium fuel fabrication plant and U.K. $^{235}$U enrichment workers at the Capenhurst facility showed reduced lung cancer rates [165, 166]. Less than expected lung cancer rates have been found in Rocky Flats plutonium workers at lung doses <400 mSv [167, 168]. PROFAC values of 80 and 86% (1 observed/7 expected) for

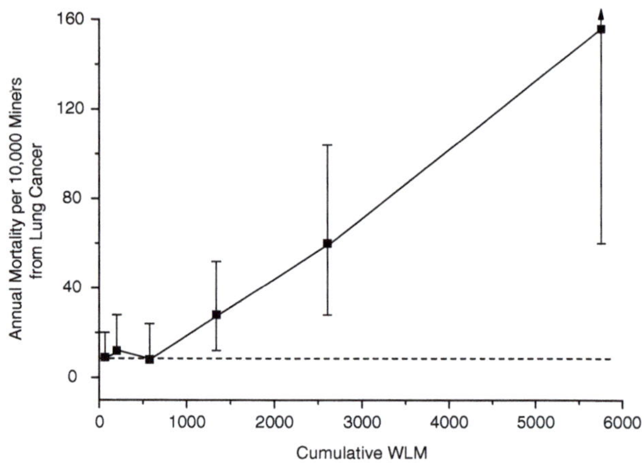


**Fig. 9.5** Lung cancer mortality among U.S. underground uranium miners [84]

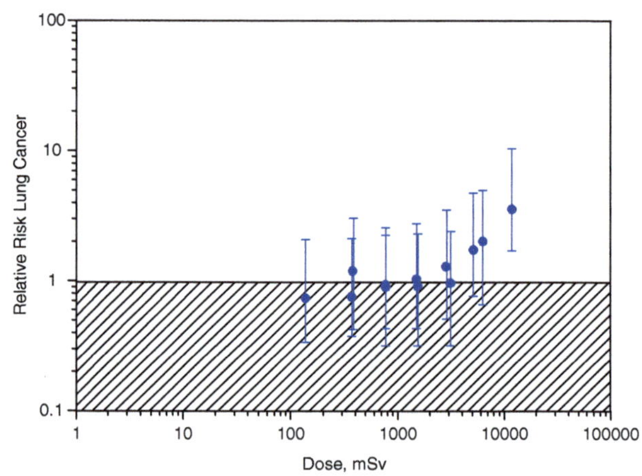


**Fig. 9.6** Relative risk of lung cancer in German uranium and Chinese tin miners [134, 158] (1 WLM = 5 mSv)

lung cancer mortality were found in plutonium workers at the Rocky Flats Plant [169, 170]. Manhattan plutonium workers had a median internal dose of 1.25 Sv and a PROFAC for lung cancer of 32% [171].

Lung tumor incidence in plutonium workers was plotted with α-radiation dose from data found in published epidemiological studies. Data were highly variable with about as many dose points showing a lung cancer RR > 1.0 as showing a RR < 1.0 at lung doses <1,000 mSv. A linear-type increased risk was seen at doses > ~1,000 mSv (Fig. 9.7).

A case–control study was carried out for all morphologically verifiable lung cancer cases from 1966 to 1991 among Mayak nuclear workers. The incidence of lung cancer at lung doses <0.8 Gy was significantly less ($p < 0.05$) than control levels (PROFACs of 44, 41, and 17% at average $^{239}$Pu body burdens of 0.34 , 1.2 , and 4.2 kBq, respectively) (Fig. 9.8) [172, 173]. An increased risk of lung cancer was found in Mayak Pu workers only at alpha doses >1,800 mSv, which were statistically significant only at doses >7,000 mSv [174]. The RR for lung cancer in Russian radiochemical workers with lung doses of ~1 Gy was 0.65; nearly all lung cancer cases were found in smokers [175]. The workers had mean lung doses that

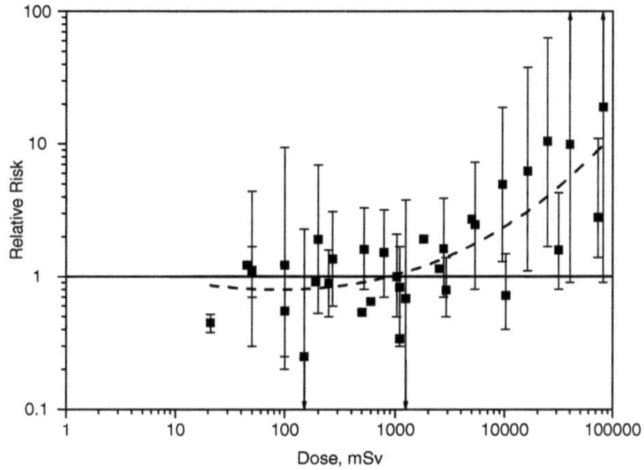


**Fig. 9.7** Lung cancer risk following exposure of the lung to high-LET radiation (mostly plutonium). The dashed curve was fitted by eye

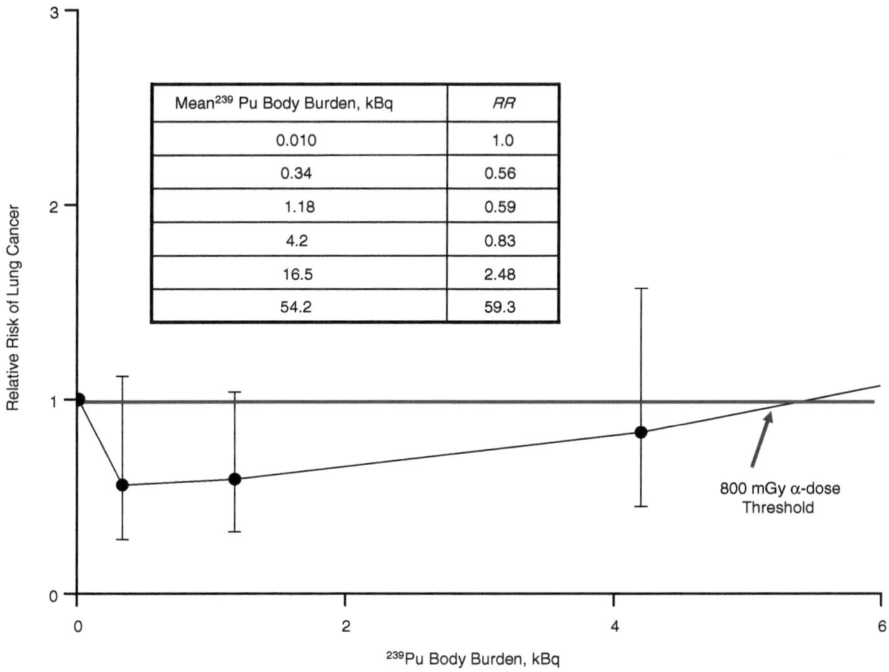

| Mean$^{239}$ Pu Body Burden, kBq | *RR* |
|---|---|
| 0.010 | 1.0 |
| 0.34 | 0.56 |
| 1.18 | 0.59 |
| 4.2 | 0.83 |
| 16.5 | 2.48 |
| 54.2 | 59.3 |



**Fig. 9.8** Relative risk of lung cancer in Mayak plutonium workers. Workers were also exposed to a mean cumulative dose of ~1 Gy $^{60}$Co γ-rays [172]

ranged from 1 to 2 Gy. Nearly all lung cancer cases were found in smokers. Significant impairment of respiratory function is caused by cigarette smoking. Respiratory function was decreased by α-irradiation from inhaled $^{239}$Pu but is not altered by whole-body gamma radiation doses up to 4 Gy [176].

The Mayak plutonium worker groups received significant low-LET γ-irradiation, mostly from $^{60}Co$. Estimates of radiation-induced lung cancer have been previously overestimated by not adjusting for cigarette smoking. A linear fit for lung cancer and smoking index has been employed for Mayak workers [172, 177, 178]. Adjusting for smoking gave a RR = 0.91 for lung cancer in Mayak workers receiving lung γ-ray doses ~1 Gy [173]. The PROFAC for lung cancer among Mayak workers from chronic alpha plus gamma irradiation was 61% at a lung dose of 0.1–12 mGy and 47% for a dose of 12.1–50 mGy [179].

## 9.9 Mechanism

A low-dose protective apoptosis-mediated (PAM) process, limiting potential cancer formation, may be activated by low-dose, low-LET gamma or X-radiations [5, 86]. Low doses of protracted low-LET γ-rays, associated with radon daughters and $^{60}Co$ in Mayak workers, are thought to trigger the hormetic response and protect against stochastic effects of pulmonary α-irradiation at lung doses <1 Gy from inhaled radon and $^{239}Pu$. Low dose, low LET γ-irradiation may have induced apoptosis of $^{239}Pu$ alpha particle-induced, genomically damaged, transformed pulmonary cells, suppressing carcinogenesis in the lung up to lung α-doses of ~5 Gy [180, 181]. Low LET radiation reduces chemical carcinogen-induced cancers in rodents [182, 183]. Low dose ionizing radiation may enhance the elimination by apoptosis of cigarette-induced transformed pulmonary cells, thus decreasing lung cancer risk [73]. The results indicated that not only did hormesis-associated protective processes prevent spontaneous cancer but also reduced the number of lung cancers associated with both α-irradiation and cigarette smoking.

## Appendix

**Table A9.6** Risk of lung cancer from epidemiologic studies of populations exposed to external low LET ionizing radiation compared with the corresponding referent groups

| Study | Dose, mSv | RR | 90–95% CI | Reference |
|---|---|---|---|---|
| UK Nuclear Test Servicemen | ~30 | 0.64 | – | Darby SC, Muirhead CR, Doll R et al (1990) Mortality among United Kingdom servicemen who served abroad in the 1950s and 1960s. Br J Ind Med 47:793–804 |
| Taiwan | 48 | 0.80 | 0.4–1.5 | Hwang SL, Guo HR, Hsieh WA et al (2006) Cancer risks in a population with prolonged low-dose-rate γ-radiation exposure in radiocontaminated buildings, 1983–2022. Int J Radiat Biol 82:849–858 |

**Table A9.6** (continued)

| Study | Dose, mSv | RR | 90–95% CI | Reference |
|---|---|---|---|---|
| Canadian Dose Registry | 32 | 0.64 | 0.59–0.70 | Sont WN, Zielinski JM, Ashmore JP et al (2001) First analysis of cancer incidence and occupational radiation exposure based on the National Dose Registry of Canada. Am J Epidemiol 153:309–318 |
| German Airlines | 100 | 0.57 | 0.06–2.28 | Blettner M, Zeeb H, Langner I et al (2002) Mortality from cancer and other causes among airline cabin attendants in Germany, 1960–1997. Am J Epidemiol 156:556–565 |
| Canadian pilots | 117 | 0.25 | 0.12–0.45 | Band PR, Le ND, Fang R et al (1996) Cohort study of Air Canada pilots: mortality, cancer incidence, and leukemia risk. Am J Epidemiol 143:137–143 |
| Airline Attendants | 110 | 0.41 | 0.08–1.21 | Reynolds P, Cone J, Layefsky M et al (2002) Cancer incidence in California flight attendants (United States). Cancer Causes Control 13:317–324 |
| US Nuclear Shipyard | 97 | 0.93 | 0.90–1.17 | Sponsler R, Cameron JR (2005) Nuclear shipyard worker study (1980–1988): a large cohort exposed to low-dose-rate gamma radiation. Int J Low Radiat 1:463–478 |
| UKAEA Workers | 19 | 0.89 | 0.82–0.98 | Atkinson WD, Law DV, Bromley KJ et al (2004) Mortality of employees of the United Kingdom Atomic Energy Authority, 1946–97. Occup Environ Med 61:577–585 |
| 12 DOE Labs (female) | 5.8 | 0.86 | 0.78–0.94 | Wilkinson GS, Trieff N, Graham R et al (2000) Final Report. Mortality among female nuclear weapons workers. NIOSH |
| US Radiological Technicians | 10<br>98<br>320 | 0.60<br>0.80<br>0.70 | –<br>–<br>– | Doody MM, Mandel JS, Lubin JH et al (1998) Mortality among United States radiologic technicians, 1926–90. Cancer Causes Control 9:67–75 |

(*continued*)

**Table A9.6** (continued)

| Study | Dose, mSv | RR | 90–95% CI | Reference |
|---|---|---|---|---|
| US Radiological Technicians | ~30 | 0.74 | 0.65–0.80 | Mohan AK, Hauptmann M, Freedman DM et al (2003) Cancer and other causes of mortality among radiologic technologists in the United States. Int J Cancer 103:259–267 |
| British Radiologists | 350 | 0.00 | – | Berrington A, Darby SC, Weiss HA et al (2001) 100 years of observation on British radiologists: mortality from cancer and other causes 1987–1997. Br J Radiol 74:507–519 |
| | 700 | 0.74 | – | |
| Three UK Nuclear Workforces | 35 | 0.98 | – | Carpenter LM, Higgins CD, Douglas DJ et al (1998) Cancer mortality in relation to monitoring for radionuclide exposure in three UK nuclear industry workforces. Br J Cancer 78:1224–1232 |
| | 75 | 0.88 | – | |
| | 200 | 0.72 | – | |
| Canadian Nuclear Power Workers | 10 | 0.69 | 0.5–1.0 | Zablotska LB, Ashmore JP, Howe GR (2004) Analysis of mortality among Canadian nuclear power industry workers after chronic low-dose exposure to ionizing radiation. Radiat Res 161:633–641 |
| | 70 | 0.74 | 0.–1.8 | |
| Australia Nuclear Workers | 15 | 0.47 | 0.28–0.76 | Habib RR, Abdallah SM, Law M, Kaldor J (2006) Cancer incidence among Australian nuclear industry workers. J Occup Health 48:358–365 |
| Rocketdyne Workers | 3 | 0.94 | – | Boice JD, Cohen SS, Mumma MT et al (2006) Mortality among radiation workers at Rocketdyne (Atomics International), 1948–1999. Radiat Res 166:98–115 |
| | 7 | 0.87 | – | |
| | 25 | 0.86 | – | |
| | 75 | 0.00 | – | |
| Portsmouth Gas Uranium | 5 | 0.78 | 0.64–0.94 | Ahrenholz S, Cardarelli J, Dill JP et al (2001) Final Report. Mortality patterns among workers at the Portsmouth gaseous diffusion plant Piketon, Ohio. NIOSH |
| INEEL Workers | 13 | 0.79 | 0.69–0.91 | Schubauer-Berigan MK, Macievic GV, Utterback DF et al (2005) Final Report. An epidemiologic study of mortality and radiation-related risk of cancer among workers at the Idaho National Engineering and Environmental Laboratory, a U.S. Department of Energy facility. NIOSH |

**Table A9.6** (continued)

| Study | Dose, mSv | RR | 90–95% CI | Reference |
|---|---|---|---|---|
| US Nuclear Power Workers | 26 | 0.59 | 0.49–0.71 | Howe GR, Zablotska LB, Fix JJ et al (2004) Analysis of the mortality experience amongst U.S. nuclear power industry workers after chronic low-dose exposure to ionizing radiation. Radiat Res 162:517–526 |
| UK Nuclear Test | 6 | 0.87 | 0.63–1.19 | Muirhead CR, Kendall GM, Darby SC et al (2004) Epidemiological studies of UK test veterans: II. Mortality and cancer incidence. J Radiol Prot 24:219–241 |
| Chapelcross, UK | 15<br>75<br>300<br>500 | 0.83<br>0.51<br>0.71<br>0.60 | –<br>–<br>–<br>– | McGeoghegan D, Binks K (2001) The mortality and cancer morbidity experience of employees at the Chapelcross plant of British Nuclear Fuels plc, 1955–95. J Radiol Prot 21:221–250 |
| Capenhurst, UK | 9.9 | 0.84 | – | McGeoghegan D, Binks K (2000) The mortality and cancer morbidity experience of workers at the Capenhurst uranium enrichment facility 1946–95. J Radiol Prot 20:381–401 |
| Springfields, UK | 9.3 | 0.99 | – | McGeoghegan D, Binks K (2000) The mortality and cancer morbidity experience of workers at the Springfields uranium production facility, 1946–95. J Radiol Prot 20:111–137 |
| Japan Nuclear Workers | 5<br>35<br>150 | 0.97<br>0.56<br>0.95 | 0.7–1.3<br>0.2–1.4<br>0.1–3.4 | Iwasaki T, Murata M, Ohshima S et al (2003) Second analysis of mortality of nuclear industry workers in Japan. Radiat Res 159:228–238 |
| Japan Nuclear Workers | 14 | 0.94 | – | Epidemiological Study Group of Nuclear Workers (Japan) (1997) First analysis of mortality of nuclear industry workers in Japan, 1986–1992. J Health Phys 32:173–184 |
| LANL Workers | 30<br>75 | 0.51<br>0.87 | –<br>– | Wiggs LD, Johnson ER, Cox-De Vore CA et al (1994) Mortality through 1990 among white male workers at the Los Alamos National Laboratory: |

(*continued*)

**Table A9.6** (continued)

| Study | Dose, mSv | RR | 90–95% CI | Reference |
|---|---|---|---|---|
| | | | | Considering exposures to plutonium and external ionizing radiation. Health Phys 67:577–588 |
| Korean Nuclear Workers | 10 | 0.99 | 0.45–1.88 | Lim YK, Yoo KY (2006) A cohort study of cancer risk by low-dose radiation exposure among radiation workers of nuclear power plants in Korea. J Korea Assoc Radiat Prot 31:53–63 |
| Naval Shipyard Workers | 75 | 0.84 | 0.46–1.44 | Yiin JH, Schubauer-Berigan MK, Silver SR et al (2005) Risk of lung cancer and leukemia from exposure to ionizing radiation and potential confounders among workers at the Portsmouth Naval Shipyard. Radiat Res 163:603–613 |
| Radiological Technologists | 5<br>48<br>150 | 0.60<br>0.80<br>0.70 | –<br>–<br>– | Simon SL, Weinstock RM, Doody MM et al (2006) Estimating historical radiation doses to a cohort of U.S. radiologic technologists. Radiat Res 166:174–192 |
| Nuclear Weapons Workers | 7 | 0.83 | 0.67–1.01 | Silver SR, Anderson-Mahoney P, Burphy J et al (2005) Mortality update for the Pantex weapons facility: final report health-related energy research branch. National Institute for Occupational Safety and Health, HHS (NIOSH) Publication No. 2005–124 |
| French Nuclear Power Workers | 18 | 0.48 | 0.33–0.69 | Rogel A, Carre N, Amoros E et al (2005) Mortality of workers exposed to ionizing radiation at the French National Electricity company. Am J Indust Med 47:72–82 |
| Radiotherapy Patients | 10 | 0.60 | 0.07–2.17 | Ron E, Auvinen A, Alfandary E et al (1999) Cancer risk following radiotherapy for infertility or menstrual disorders. Int J Cancer 82:795–798 |
| French AEC Workers | ~10 | 0.79 | 0.58–1.08 | Telle-Lamberton M, Bergot D, Gagneau M et al (2004) Cancer mortality among French Atomic Energy Commission workers. Am J Indust Med 45:34–44 |

**Table A9.6** (continued)

| Study | Dose, mSv | RR | 90–95% CI | Reference |
|---|---|---|---|---|
| Russian Nuclear Workers | 70 | 0.52 | 0.31–0.82 | Ivanov V, Iiyin L, Gorski A et al (2004) Radiation and epidemiological analysis for solid cancer incidence among nuclear workers who participated in recovery operations following the accident at the Chernobyl NPP. J Radiat Res (Tokyo) 45:41–44 |
| Dose Registry of Canada | 6.3 | 0.63 | 0.53–074 | Ashmore JP, Krewski D, Ziellnski JM et al (1998) First analysis of mortality and occupational radiation exposure based on the National Dose Registry of Canada. Am J Epidemiol 148:564–574 |
| Registry Nuclear Workers | 3.2 | 0.98 | – | Muirhead CR, Goodill AA, Haylock RGE et al (1999) Occupational radiation exposure and mortality: second analysis of the National Registry of Radiation Workers. J Radiol Prot 19:3–26 |
| | 14 | 0.93 | – | |
| | 30 | 0.71 | – | |
| | 71 | 0.82 | – | |
| | 172 | 0.54 | – | |
| Canadian TB Patients | 250 | 0.87 | 0.7–1.0 | Howe GR (1995) Lung cancer mortality between 1950 and 1987 after exposure to fractionated moderate-dose-rate ionizing radiation in the Canadian fluoroscopy cohort study and a comparison with lung cancer mortality in the Atomic Bomb survivors study. Radiat Res 142:295–304 |
| | 750 | 0.82 | 0.7–1.0 | |
| | 1,500 | 0.94 | 0.8–1.2 | |
| Massachusetts TB Patients | 840 | 0.84 | – | Davis F, Boice J, Hrubec Z et al (1989) Lung cancer mortality in a radiation-exposed cohort of Massachusetts tuberculosis patients. Cancer Res 49:6130–6136 |
| Residents Yangjiang, China | 200 | 0.81 | 0.53–1.24 | Tao T, Zha Y, Akiba S et al (2000) Cancer mortality in the high background radiation areas of Yangjiang, China during the period between 1979 and 1995. J Radiat Res 41(Suppl):31–41 |
| Radioiodine Therapy Patients | 280 | 0.76 | 0.61–0.95 | Franklyn JA, Maisonneuve P, Sheppard M et al (1999) Cancer incidence and mortality after radioiodine treatment for hyperthyroidism: a population-based cohort study. Lancet 353:2111–2115 |

*(continued)*

**Table A9.6** (continued)

| Study | Dose, mSv | RR | 90–95% CI | Reference |
|---|---|---|---|---|
| Mayak Pu Workers (external) | 250 | 0.56 | 0.3–1.2 | Koshurnikova NA, Bolotnikova MG, Iyin LA et al (1998) Lung cancer risk due to exposure to incorporated plutonium. Radiat Res 149:366–371 |
| | 750 | 0.84 | 0.4–1.8 | |
| | 3,000 | 0.72 | 0.3–2.0 | |

**Table A9.7** Case–Control Studies of Indoor Radon and Lung Cancer

| Study | Dose | RR | 95% CI | Reference |
|---|---|---|---|---|
| Italy | 0–49 Bq m$^{-3}$ | 1.0 | 0.57–1.76 | Bochicchio F, Forastiere F, Faqrchi S et al (2005) Residential radon exposure, diet and lung cancer: a case-control study in a Mediterranean region. Int J Cancer 114:983–991 |
| | 50–99 | 1.28 | 1.02–1.60 | |
| | 100–199 | 1.45 | 1.08–1.96 | |
| | 200–399 | 1.27 | 0.72–2.25 | |
| | 400+ | 2.97 | 0.46–18.9 | |
| Missouri, USA | <30 | 1.00 | | Alavanja MCR, Brownson RC, Lubin JH et al (1994) Residential radon exposure and lung cancer among nonsmoking women. JNCI 86:1829–1837 |
| | 30–43 | 1.01 | 0.7–1.4 | |
| | 44–63 | 0.84 | 0.6–1.3 | |
| | 64–91 | 0.90 | 0.6–1.3 | |
| | >91 | 1.2 | 0.9–1.7 | |
| Iowa, USA | <57 | 1.00 | | Field RW, Steck DJ, Smith BJ et al (2000) Residential radon gas exposure and lung cancer. The Iowa Radon Lung Cancer Study. Am J Epidemiol 151:1091–1102 |
| | 58–114 | 1.34 | 0.81–2.22 | |
| | 115–170 | 1.73 | 0.99–3.04 | |
| | 171–228 | 1.62 | 0.88–2.99 | |
| | >228 | 1.79 | 0.99–3.26 | |
| Gansu, China | <100 Bq m$^{-3}$ | 1.00 | | Wang Z, Lubin JH, Wang L et al (2002) Residential radon and lung cancer risk in a high-exposure area of Gansu Province, China. Am J epidemiol 155:554–564 |
| | 100–149 | 1.00 | 0.7–1.5 | |
| | 150–199 | 1.42 | 1.0–2.0 | |
| | 200–249 | 1.36 | 1.0–1.9 | |
| | 250–299 | 1.28 | 0.8–1.9 | |
| | >300 | 1.58 | 1.1–2.3 | |
| Western Germany | 0–50 Bq m$^{-3}$ | 1.0 | | Kreienbrock L, Kreuzer M, Gerken M et al (2001) Case-control study on lung cancer and residential radon in western Germany. Am J Epidemiol 153:42–52 |
| | 50–80 | 0.98 | 0.81–1.20 | |
| | 80–140 | 1.09 | 0.80–1.48 | |
| | >140 | 0.99 | 0.61–1.63 | |
| Nationwide Swedish Study | <50 Bq m$^{-3}$ | 1.0 | | Lagarde F, Axelsson G, Damber L et al (2001) Residential radon and lung cancer among never-smokers in Sweden. Epidemiology 12:396–404 |
| | 50–80 | 1.08 | 0.79–1.47 | |
| | 80–140 | 1.18 | 0.86–1.61 | |
| | >140 | 1.44 | 1.00–2.06 | |
| Spain | 0–36.9 Bq m$^{-3}$ | 1.00 | | Barros-Dios JM, Barreiro MA, Ruano-Ravina A, Figueiras A (2002) Exposure to residential radon and lung cancer in Spain: a population-based case-control study. Am J Epidemiol 156:548–555 |
| | 37–55.1 | 2.73 | 1.12–5.48 | |
| | 55.2–147.9 | 2.48 | 1.29–6.79 | |
| | >148 | 2.96 | 1.29–6.79 | |

**Table A9.7** (continued)

| Study | Dose | RR | 95% CI | Reference |
|---|---|---|---|---|
| Shenyang, China | <100 | 1.0 | | Lubin JH, Wang ZY, Boice JD et al (2003) Risk of lung cancer and residential radon in China: pooled results of two studies. Int J Cancer 109:132–137 |
| | 100–149 | 1.29 | 0.8–2.0 | |
| | 150–199 | 1.16 | 0.5–2.7 | |
| | 200–249 | 0.75 | 0.3–2.0 | |
| | 250–299 | 1.06 | 0.2–5.8 | |
| | >300 | 1.05 | 0.4–3.0 | |
| France | <50 Bq m$^{-3}$ | 1.0 | | Baysson H, Tirmarche M, Tymen G et al (2004) Indoor radon and lung cancer in France. Epidemiology 15:709–716 |
| | 50–100 | 0.85 | 0.59–1.22 | |
| | 100–200 | 1.19 | 0.81–1.77 | |
| | 200–400 | 1.04 | 0.64–1.67 | |
| | >400 | 1.11 | 0.59–2.09 | |
| Finland | <50 | 1.0 | | Auvinen A, Makelainen H, Hakama M et al (1996) Indoor radon exposure and risk of lung cancer: a nested case-control study in Finland. JNCI 88:966–972 |
| | 50–99 | 1.03 | 0.84–1.26 | |
| | 100–199 | 1.00 | 0.78–1.29 | |
| | 200–399 | 0. 91 | 0.61–1.35 | |
| | 400–1,277 | 1.15 | 0.69–1.93 | |
| Germany | <50 Bq m$^{-3}$ | 1.00 | | Wichmann H-E, Schaffrath Rosario A, Heid IM et al (2005) Increased lung cancer risk due to residential radon in a pooled and extended analysis of studies in Germany. Health Phys 88:71–79 |
| | 50–80 | 0.97 | 0.85–1.11 | |
| | 80–140 | 1.06 | 0.87–1.30 | |
| | >140 | 1.40 | 1.03–1.89 | |
| Misasa, Japan | <25 Bq m$^{-3}$ | 1.00 | | Sobue T, Lee VS, Ye W et al (2000) Residential radon exposure and lung cancer risk in Misasa, Japan: a case-control study. J Radiat Res 41:81–92 |
| | 25–49 | 1.13 | 0.29–4.40 | |
| | 50–99 | 1.23 | 0.16–9.39 | |
| | 100 + | 0.25 | 0.03–2.33 | |
| Finland | <95 Bq m$^{-3}$ | 1.00 | | Ruosteenoja E, Makelainen L, Tapio H et al (1996) Radon and Lung Cancer in Finland. Health Phys. 71:185–189 |
| | 95–186 | 1.8 | | |
| | >186 | 1.5 | | |
| Italy | <40 | 1.00 | | Pisa FE, Barbone F, Betta A et al (2001) Residential radon and risk of lung cancer in an Italian alpine area. Arch Environ Health 56:208–215 |
| | 40–76 | 2.00 | 1.00–3.9 | |
| | 77–139 | 1.8 | 0.9–6.2 | |
| | 140–199 | 2.4 | 0.9–6.2 | |
| | >200 | 1.0 | 0.3–3.1 | |
| China | <74 | 1.00 | | Blot WJ, Xu ZY, Boice JD et al (1990) Indoor radon and lung cancer in China. JNCI 82:1025–1030 |
| | 74–147 | 0.9 | 0.6–1.3 | |
| | 148–295 | 0.9 | 0.5–1.4 | |
| | >295 | 0.7 | 0.4–1.3 | |
| New Jersey | <37 | 1.00 | | Schenberg JB, Klotz JB, Wilcox HB et al (1992) A case-control study of radon and lung cancer among New Jersey women. In: Cross FT (ed) Indoor radon and lung cancer: reality or myth? Twenty-ninth Hanford Symposium on Health and the Environment, Battelle Press, Columbus, OH, pp 905–918 |
| | 37–73 | 1.2 | 0.83–1.7 | |
| | 74–147 | 1.1 | 0.55–2.3 | |

*(continued)*

**Table A9.7** (continued)

| Study | Dose | RR | 95% CI | Reference |
|---|---|---|---|---|
| Sweden | <75 | 1.00 | | Pershagen G, Liang ZH, Hrubec Z et al (1992) Residential radon exposure and lung cancedr in Swedish women. Health Phys 63:179–186 |
| | 75–110 | 1.2 | 0.7–1.4 | |
| | 111–150 | 1.3 | 0.7–2.3 | |
| | >150 | 1.7 | 1.0–2.4 | |
| SW England | <25 | 1.0 | | Darby S, Whitley E, Silcocks P et al (1998) Risk of lung cancer associated with residential radon exposure in south-west England: a case-control study. Br J Cancer 78:394–408 |
| | 25–49 | 1.06 | 0.88–1.29 | |
| | 50–99 | 1.13 | 0.89–1.44 | |
| | 100–199 | 1.29 | 0.79–2.12 | |
| | 200–399 | 0.94 | 0.68–1.29 | |
| | >400 | 1.79 | 0.74–4.33 | |
| Missouri | <37 | 1.00 | | Alavanja MCR, Lubin JH, Mahaffey JA et al (1999) Residential radon exposure and risk of lung cancer in Missouri. Am J Public Health 89:1042–1048 |
| | 37–73 | 0.87 | 0.6–1.3 | |
| | 74–147 | 0.91 | 0.5–1.5 | |
| | >148 | 0.71 | 0.3–1.3 | |

## References


1. Field AW, D Krewski, JH Lubin et al (2006) An overview of the North American residential radon and lung cancer case-control studies. J Toxicol Environ Hlth Part A 69:599–631
2. The health benefits of smoking cessation: a report of the Surgeon General. DHHS publication no. (CDC) 90-8416 (1990) Department of Health and Human Services, Rockville, MD
3. Jemal A, A Thomas, T Murray et al (2002) Cancer statistics. CA Cancer J Clin 52:23–47
4. Sanders CL (1996) Prevention and therapy of cancer and other common diseases: alternative and traditional approaches. Infomedix, Richland, WA, 3000 pp
5. Sanders CL, BR Scott (2008) Smoking and hormesis as confounding factors in radiation pulmonary carcinogenesis. Dose Response 6:53–79
6. Bochicchio F, F Forastiere, S Farchi et al (2005) Residential radon exposure, diet and lung cancer: a case-control study in a Mediterranean region. Int J Cancer 114:983–991
7. Blot WJ (1992) Alcohol and cancer. Cancer Res 52(Suppl):2119s–2123s
8. Hammond EC, IJ Selikoff, H Seidman H (1979) Asbestos exposure, cigarette smoking and death rates. Ann NY Acad Sci 330:473–491
9. U.S. Environmental Protection Agency (1992) Respiratory health effects of passive smoking: lung cancers and other disorders. EPA/600//-90/006F, Washington, DC
10. Fontham ETH, P Correa, P Reynolds et al (1994) Environmental tobacco smoke and lung cancer in nonsmoking women. J Amer Med Assoc 271:1752–1759
11. Lave LB, EP Seskin (1970) Air pollution and human health. Science 169:723–733
12. Desideri D, MA Meli, L Feduzi et al (2007) $^{210}$Po and $^{210}$Pb inhalation by cigarette smoking in Italy. Health Phys 92:58–63
13. Thompson TG (2004) The Health Consequences of Smoking: A Report of the Surgeon General, www.cdc.gov/tobacco/sgr/.
14. Deetjen P, Falkenbach A (eds) (1999) Radon und Gesundheit. P. Land, Frankfurt, Germany
15. Chyou PH, AM Nomura, GN Stemmermann (1992) A prospective study of the attributable risk of cancer due to cigarette smoking. Amer J Public Health 82:37–40
16. Sethi T (1997) Science, medicine, and the future: lung cancer. Brit J Med 314:652–660

17. U. S. Public Health Service (1982) The health consequences of smoking: Cancer. A report of the Surgeon General. U. S. Publich Health Service, DHHS Publication no. (PHS) 82–50179.
18. Crispo A, P Brennan, KH Jockel et al (2004) The cumulative risk of lung cancer among current, ex- and never-smokers in European men. Br J Cancer 91:1280–1286; Vineis P, M Alavanja, P Buffler, et al. 2004. Tobacco and cancer: recent epidemiological evidence. J Natl Cancer Inst 96:99–106.
19. Heidenreich WF, EG Luebeck, WD Hazelton et al (2002) Mutlistage models and the incidence of cancer in the cohort of atomic bomb survivors. Radiat Res 158:607–614
20. Gustavsson P, R Jakobsson, F Nyberg et al (2000) Occupational exposure and lung cancer risk: a population-based case-referent study in Sweden. Am J Epidemiol 152:32–40
21. IARC (International Agency for Research on Cancer) (1987) Overall evaluations of carcinogenicity: an updating of IARC monographs volumes 1 to 42. (IARC monographs on the evaluation of carcinogenic risks to humans, supplement no. 7). IARC, Lyon, France, pp 1–440
22. Toth E, Lazar I, Selmeczi D et al (1998) Lower cancer risk in medium high radon. Pathol Oncol Res 4:125–129
23. Villeneuve PJ, HI Morrison, R Lane (2007) Radon and lung cancer risk: an extension of the mortality follow-up of the Newfoundland fluorspar cohort. Health Phys 92:157–169
24. Doll R (1992) The lessons of life: keynote address to the nutrition and cancer conference. Cancer Res 52:2024s–2029s
25. Schull WJ, KM Weiss (1992) Radiation carcinogenesis in humans. Adv Radiat Biol 16: 215–258
26. Gilbert ES, NA Koshurnikova, ME Sokolnikov et al (2004) Lung cancer in Mayak workers. Radiat Res 162:505–516
27. Kreisheimer M, NA Koshurnikova, E Nekolla et al (2000) Lung cancer mortality among male nuclear workers of the Mayak facilities in the former Soviet Union. Radiat Res 154:3–11
28. Shilnikova NS, DL Preston, E Ron et al (2003) Cancer mortality risk among workers at the Mayak nuclear complex. Radiat Res 159:787–798
29. Yiin JH, MK Schubauer-Berigan, SR Silver et al (2005) Risk of lung cancer and leukemia from exposure to ionizing radiation and potential confounders among workers at the Portsmouth Naval Shipyard. Radiat Res 163:603–613
30. Cardis E, M Vrijheid, M Blettner et al (2005) Risk of cancer after low doses of ionizing radiation: retrospective cohort study in 15 countries. Brit Med J 331:77–80
31. Pierce DA, GB Sharp, K Mabuchi (2003) Joint effects of radiation and smoking on lung cancer risk among atomic bomb survivors. Radiat Res 159:511–520
32. Pierce DA, Y Shimizu, DL Preston et al (1996) Studies of the mortality of atomic bomb survivors. Report 12. 1. Cancer: 1950–1990. Radiat Res 146:1–27
33. Kopecky KJ, E Nakashima, T Yamamoto et al (1986) Lung cancer, radiation, and smoking among A-bomb survivors, Hiroshima and Nagasaki: TR-13–86. Radiation Effects Research Foundation, Hiroshima
34. Little MP (2002) Comparisons of lung tumour mortality risk in the Japanese A-bomb survivors and in the Colorado Plateau uranium miners: support for the ICRP lung model. Intern J Radiat Biol 78:145–163
35. Taylor LS (1980) Some non-scientific influences on radiation protection standards and practice. Health Phys 32:851–874
36. Cameron JR (2002) Correspondence: Radiation increased the longevity of British radiologists. Br J Radiol 75:637–639
37. Keirim-Markus IB (2004) Radiation exposure normalization taking account of specific effects at low doses and dose rates. Atomic Energy 93:836–844
38. Shleien B, AJ Ruttenber, M Sage (1991) Epidemiologic studies of cancer in populations near nuclear facilities. Health Phys 61:699–713
39. Atkinson WD, DV Law, KJ Bromley et al (2004) Mortality of employees of the United Kingdom Atomic Energy Authority, 1946–97. Occup Environ Med 61:577–585

40. Ivanov V, L Iiyin, A Gorski et al (2004) Radiation and epidemiological analysis for solid cancer incidence among nuclear workers who participated in recovery operations following the accident at the Chernobyl NPP. J Radiat Res (Tokyo) 45:41–44
41. Hornung RW, TJ Meinhardt (1987) Quantitative risk assessment of lung cancer in U. S. uranium miners. Health Phys 52:417–430
42. Enflo A (2002) Lung cancer risks from residential radon among smokers and non-smokers. J Radiol Prot 22:A95–A99
43. Petersen GR, ES Gilbert, JA Buchanan et al (1989) A case-cohort study of lung cancer, ionizing radiation, and tobacco smoking among males at the Hanford site. Health Phys 58:3–11
44. Blettner M, H Zeeb, I Langner et al (2002) Mortality from cancer and other causes among airline cabin attendants in Germany, 1960–1997. Am J Epidemiol 156:556–565
45. Band PR, ND Le, R Fang et al (1996) Cohort study of Air Canada pilots: mortality, cancer incidence, and leukemia risk. Am J Epidemiol 143:137–143
46. Baverstock K, D Williams (2002) Chernobyl: an overlooked aspect? Letters Science 299:44
47. Chernobyl Forum (IAEA, WHO, UNDP, UNEP, UN-OCHA, UNSCEAR, World Bank) (2005) Chernobyl's Legacy: Health, Environmental and Socio-Economic Impacts. The work is in three volumes and 600 pages by more than 100 scientists
48. Hatch M, E Ron, A Bouville et al (2005) The Chernobyl disaster: cancer following the accident at the Chernobyl nuclear power plant. Epidemiol Reviews 27:56–66
49. Rossi HH, M Zaider (1997) Radiogenic lung cancer: the effects of low doses of low linear energy transfer (LET) radiation. Radiat Environ Biophys 36:85–88
50. Thompson RC, JA Mahaffey (1986) Life-span radiation effects studies in animals: what can they tell us? Proceedings of the 22nd Hanford Life Science Symposium, Richland, WA
51. Stannard JN (1988) Radioactivity and health: a history. In: Baalman RW (ed). Pacific Northwest Laboratory, Richland, WA
52. Gregoire O, MR Cleland (2006) Novel approach to analyzing the carcinogenic effect of ionizing radiations. Int J Radiat Biol 82:13–19
53. Okamoto K (1987) Critical values of linear energy transfer, dose rates and doses for radiation hormesis. Health Phys 52:671–674
54. Wilkinson GS, Trieff N, Graham R et al (2000) Final Report. Study of Mortality Among Female Nuclear Weapons Workers. Grant Numbers: 1R01 OHO3274, R01/CCR214546, R01/CCR61 2934-01, National Institute for Occupational Safety and Health, Centers for Disease Control and Prevention
55. Matanoski GM (1991) Health effects of low-level radiation in shipyard workers. Final Report. Report No. DOE DE-AC02-79EV10095. US Department of Energy, Washington, DC
56. Sont WN, JM Zielinski, AP Ashmore et al (2001) First analysis of cancer incidence and occupational radiation exposure based on the National Dose Registry of Canada. Am J Epidemiol 153:309–318
57. Zablotska LB, Ashmore JP, Howe GR (2004) Analysis of mortality among Canadian nuclear power industry workers after chronic low-dose exposure to ionizing radiation. Radiat Res 161:633–641
58. McGeoghegan D, K Binks (2001) The mortality and cancer morbidity experience of employees at the Chapelcross plant of British Nuclear Fuels, 1955–95. J Radiol Prot 21:221–250
59. Carpenter LM, V Beral, PG Smith (1998) Cancer mortality in relation to monitoring for radionuclide exposure in three UK nuclear industry workforces. Br J Cancer 78:1224–1232
60. Iwasaki T, M Murata, S Ohshima et al (2003) Second analysis of nuclear industry workers in Japan, 1986–1997. Radiat Res 159:228–238
61. Ahn Y, J Bae (2005) Chronic exposure of low-dose radiation and cancer risks among nuclear power plant workers in Korea. Proceedings of the Forty-Eighth Annual Meeting of the Japan Radiation Research Society/The First Asian Congress of Radiation Research, Research Institute for Radiation Biology and Medicine, Hiroshima University, Japan, Abstract S11-2, p 89

62. Jin YW, M Jeong, SH Sung et al (2002) Epidemiological investigation of deaths among radiation workers in nuclear power plants of Korea. J Korea Assoc Radiat Prot 27:233–237
63. Howe GR, LB Zablotska, JJ Fix et al (2004) Analysis of the mortality experience amongst U.S. nuclear power industry workers after chronic low-dose exposure to ionizing radiation. Radiat Res 162:517–526
64. Howe GR (1995) Lung cancer mortality between 1950 and 1987 after exposure to fractionated moderate-dose-rate ionizing radiation in the Canadian Fluoroscopy Cohort Study and a comparison with lung cancer mortality in the Atomic Bomb Survivors Study. Radiat Res 142:295–304
65. Davis FG, JD Boice, H Hrubec et al (1989) Lung cancer mortality in a radiation-exposed cohort of Massachusetts tuberculosis patients. Cancer Res 49:6130–6136
66. Luckey TD (2003) A Rosetta stone for ionizing radiation. ROS Magazine 8:22–30; lung cancer in Sweden. N Engl J Med 330:159–164
67. Berrington A, SC Darby, HA Weiss et al (2001) 100 years of observation on British radiologists: mortality from cancer and other causes 1987–1997. Br J Radiol 74:507–519
68. Doody MM, JS Mandel, JH Lubin et al (1998) Mortality among USA radiologic technologists 1926–1990. Cancer Causes Control 9:67–75
69. Yoshinaga S, T Aoyama, Y Yoshimoto et al (1999) Cancer mortality among radiological technologists in Japan: updated analysis of follow-up data from 1969 to 1993. J Epidemiol 9: 61–72
70. Sanders CL (2006) Hormesis as a confounding factor in epidemiological studies of radiation carcinogenesis. Korean Assoc Radiat Prot 31:69–89
71. Baysson H, M Tirmarche, G Tymen et al (2004) Indoor radon and lung cancer in France. Epidemiology 15:709–716
72. Krewski D, JH Lubin, JM Zielinski et al (2005) Residential radon and risk of lung cancer. A combined analysis of 7 North American case-control studies. Epidemiology 16:137–145
73. Sanders CL (2008) Prevention of cigarette smoke induced lung cancer by low LET ionizing radiation. Nuclear Engineering and Technol 40:539–550
74. Crump KS (2006) The effect of random error in exposure measurement upon the shape of the exposure response. Dose-Response 3:456–464
75. Lubin JH, JD Boice, C Edling et al (1995) Lung cancer in radon-exposed miners and estimation of risk from indoor exposure. J Natl Cancer Inst 87:817–827
76. Lubin JH, L Tomasek, C Edling et al (1997) Estimating lung cancer mortality from residential radon using data for low exposures of miners. Radiat Res 147:126–134
77. Pavia M, A Bianco, C Pileggi, IF Angelillo (2003) Meta-analysis of residential exposure to radon gas and lung cancer. Bull World Health Organization 81:732–738
78. Yarmoshenko IV, IA Kirdin, MV Zhukovsky et al (2003) Meta analysis of epidemiological case control studies of lung cancer risk and indoor radon exposure. Med Radiol Radiat Prot 48:33–43 (in Russian)
79. Darby S, D Hill, A Auvinen et al (2005) Radon in homes and risk of lung cancer: collaborative analysis of individual data from 13 European case-control studies. Br Med J 330:223–228
80. Lubin JH, JD Boice (1997) Lung cancer risk from residential radon: meta-analysis of eight epidemiological studies. J Natl Cancer Inst 89:49–57
81. Lubin JH, ZY Wang, JD Boice et al (2004) Risk of lung cancer and residential radon in China: Pooled results of two studies. Int J Cancer 109:132–137
82. Neuberger JS, TF Gesell (2002). Residential radon exposure and lung cancer: risk in nonsmokers. Health Phys 83:1–18
83. Neuberger JS, RW Field (2003) Occupation and lung cancer in nonsmokers. Rev Environ Health 18:251–267
84. National Research Council (1999) Committee on Health Risks of Exposure to Radon, Health Effects of Exposure to Radon (BEIR VI), National Academy Press, Washington, DC

85. Thomas DC, KG McNeill, C Dougherty (1985) Estimates of lifetime lung-cancer risks resulting from Rn progeny exposures. Health Phys 49:825–846
86. Scott BR, SA Belinsky, S Leng et al (2009) Radiation-stimulated epigenetic reprogramming of adaptive-response genes in the lung-an evolutionary gift for mounting adaptive protection against lung cancer. Dose Response 7:104–131
87. Luxin WEI, T Sugahara, Z Tao (1997) High level of natural radiation 1966: radiation dose and health effects. Elsevier, Amsterdam
88. Lubin JH (1994) Invited commentary: lung cancer and exposure to residential radon. Am J Epidemiol 140:323–332
89. NCRP (National Council on Radiation Protection and Measurements) (1987) Exposure of the population in the United States and Canada from natural background radiation. NCRP Report No. 94, Bethesda, MD
90. Thompson RE, DF Nelson, JH Popkin et al (2008) Case-control study of lung cancer risk from residential radon exposure in Worcester County, Massachusetts. Health Phys 94: 228–241
91. Mitsunobu F, K Yamaoka, K Hanamoto et al (2003) Elevation of antioxidant enzymes in the clinical effects of radon and thermal therapy for bronchial asthma. J Radiat Res 44:95–99
92. Yamaoka K, C Sugie, S Futatsugawa et al (2005) Basic study of biologic effects of Thoron hot spring on hypertension. Proceedings of the Forty-Eighth Annual Meeting of the Japan Radiation Research Society/ the First Asian Congress of Radiation Research, Research Institute for Radiation Biology and Medicine, Hiroshima University, Japan. Abstract P-A-108, p 139
93. Takahashi M, S Kojima (2006) Suppression of atopic dermatitis and tumor metastasis in mice by small amounts of radon. Radiat Res 165:337–342
94. Franke A, L Reiner, P Pratzel et al (2000) Long-term efficacy of radon spa therapy in rheumatoid arthritis - a randomized, sham-controlled study and follow-up. Rheumatology 39: 894–902
95. Bogoljubov WM (1988) Clinical aspects of radon therapy in the USSR. Z Phys Med Balneol Med Klimatol 17:58–63
96. City of Fort Collins, CO (2005). Air Quality Department, http://www.ci.fort-collins.co.us/airqulaity/radon-health
97. Hickey RJ, EJ Bowers, DE Spence et al (1981) Low level ionizing radiation and human mortality: multi-regional epidemiological studies. Health Phys 40:625–641
98. Frigerio NA, RS Stowe (1976) Carcinogenic and genetic hazards from background radiation, vol 2, IAEA, Vienna, pp 385–393
99. Nambi KSV, SD Soman (1987) Environmental radiation and cancer in India. Health Phys 52:653–657
100. Sun Q, ZS Akiba, Z Tao et al (2000) Excess risk of solid cancer mortality after prolonged exposure to naturally occurring high background radiation in Yangjiang, China. J Radiat Res 41(Suppl):43–52
101. Tao T, Y Zha, S Akiba et al (2000) Cancer mortality in the high background radiation areas of Yangjiang, China during the period between 1979 and 1995. J Radiat Res 41(Suppl): 31–41)
102. Auvenin A, Maekelaeinen I, Hakama M et al (1996) Indoor radon exposure and risk of lung cancer: a nested case-control study in Finland. J Natl Cancer Inst 88:966–972
103. Letourneau EG, Y Mao, RG McGregor et al (1983) Lung cancer mortality and indoor radon concentrations in 18 Canadian cities. In: Proceedings of the Sixteenth Midyear Topical Meeting of the Health Physics Society. National Technical Information Service, CONF-830101, UC-41, Albuquerque, NM, pp 470–438
104. Mortazavi SMJ, T Ikushima (2006) Open questions regarding implications of radioadaptive response in the estimation of the risks of low-level exposures in nuclear workers. Int J Low Radiat 2:88–96

105. Neuberger JS (1992) Residential radon exposure and lung cancer: an overview of ongoing studies. Health Phys 63:503–509
106. Wang Z, JH Lubin, L Wang et al (2002) Residential radon and lung cancer risk in a high-exposure area of Gansu Province, China. Am J Epidemiol 155:554–564
107. Ghiassi-Nejad M, M Mortazavi (2005) Radiation adaptive response observed in residents in high level natural area of Ramsar. Proceedings of the Forty-Eighth Annual Meeting of the Japan Radiation Research Society/ the First Asian Congress of Radiation Research, Research Institute for Radiation Biology and Medicine, Hiroshima University, Japan. Abstract S4-2-2, p 81
108. Bowie C, SHU Bowie (1991) Radon and health. Lancet 337:409–413
109. Jagger J (1998) Natural background radiation and cancer death in Rocky Mountain states and Gulf Coast states. Health Phys 75:428–430
110. Wei L, T Sugahara (2000) An introductory overview of the epidemiological study on the population at the high background radiation areas in Yangjiang, China. J Radiat Res (Toyko) 41(Suppl):1–7
111. National Council on Radiation Protection and Measurements (2001) Evaluation of the linear-nonthreshold model for ionizing radiation. NCRP Report 136, Bethesda, MD: National Council on Radiation Protection and Measurements, p 6
112. Haynes RM (1988) The distribution of domestic radon concentrations and lung cancer mortality in England and Wales. Radiat Prot Dosim 25:93–96
113. Arndt D (1992) Die Strahlenexposition in den Bergbaugebieten Sachsens und Thüringens. In: Reiners CHR et al (ed) Strahlenschutz in Forschung und Praxis, vol 33, Stuttgart, 47–60
114. Kreienbrock L, M Kreuzer, M Gerken et al (2001) Case-control study on lung cancer and residential radon in western Germany. Am J Epidemiol 153:42–52
115. Sobue T, VS Lee, W Ye et al (2000) Residential radon exposure and lung cancer risk in Misasa, Japan: a case-control study. J Radiat Res 41:81–92
116. Cohen BL (1995) Test of the linear no-threshold theory of radiation carcinogenesis for inhaled radon decay products. Health Phys 68:157–174
117. Cohen B (1990) A test of the linear no-threshold theory of radiation carcinogenesis. Environ Res 53:193–220
118. Cohen BL (1987) Tests of the linear, no-threshold dose-response relationship for high-LET radiation. Health Phys 52:629–636
119. Cohen BL (1993) Test of the linear no-threshold theory of radiation carcinogenesis for inhaled radon decay products. Health Phys 56:154–174
120. Cohen BL (2000) Updates and extensions to tests of the linear no-threshold theory. Technology 7:657–672
121. Mao Y, J Hu, A-M Ugnat et al (2001) Socioeconomic status and lung cancer risk in Canada. Int J Epidemiol 30:809–817
122. Cohen BL (2001) Radon exposure and the risk of lung cancer. J Radiat Prot 21:64–66
123. Puskin JS (2003) Smoking as a confounder in ecologic correlations of cancer mortality rates with average county radon levels. Health Phys 84:526–532
124. Cohen BL (2004) Response to "residential radon exposures and lung cancer risk: Commentary on Cohen's county-based study". Health Phys 87:656–658
125. Stidley CA, JM Samet (1993) A review of ecologic studies of lung cancer and indoor radon. Health Phys 65:234–251
126. Van Pelt WR (2003) Epidemiological associations among lung cancer, radon exposure and elevation above sea level-a reassessment of cohen's county level radon study. Health Phys 85:397–403
127. Huff RL, JH Lawrence et al (1951) Effects of changes in altitude on hematopoietic activity. Medicine 30:197–217
128. Heath CW, PD Bond, DG Hoel et al (2004) Residential radon exposure and lung cancer risk: commentary on Cohen's county-based study. Health Phys 87:647–655

129. Price P (2004) Assessing uncertainties in the relationship between inhaled particle concentrations, internal deposition, and health effects. In: Ruzer LS (ed) Aerosol handbook: measurement, dosimetry and health effects. CRC-Taylor & Francis, pp 157–188
130. Lachet B (2005) Indoor radon and lung cancer. Letters to the Editor. Epidemiol 17:121
131. Cavallo A, A Hutter, P Shebel (1999) Radon progeny unattached fraction in an atmosphere far from radioactive equilibrium. Health Phys 76:532–536
132. James AC, A Birchall A, GH Akabani (2002) Comparative dosimetry of BEIR VI revisited. Radiat Prot Dosim 108:3–26
133. Puskin JS, AC James (2006) Radon exposure assessment and dosimetry applied to epidemiology and risk estimation. Radiat Res 166:193–208
134. Bruske-Hohfeld I, AS Rosario, G Wolke et al (2006) Lung cancer risk among former uranium miners of the Wismut Company in Germany. Health Phys 90:208–216
135. Cothern CR (1999) Indoor air radon. Environ Geochem Health 21:83–90
136. Lin CY, A Gelman, PN Price et al (1999) Analysis of local decisions using hierarchial modeling, applied to home radon radon measurement and remediation. Statistical Science 14:305–337
137. Blot WJ, Z-Y Xu, JD Boice et al (1990) Indoor radon and lung cancer in China. J Natl Cancer Inst 82:10–25
138. Kreuzer M, J Heinrich, G Wolke et al (2003) Residential radon and risk of lung cancer in Eastern Germany. Epidemiology 14:559–568
139. Letourneau EG, D Krewski, NW Choi et al (1994) Case-control study of residential radon and lung cancer in Winnipeg, Manitoba, Canada. Am J Epidemiol 140:310–322
140. Krewski D, SN Rai, JM Zielinski et al (1999) Characterization of uncertainty and variability in residential radon cancer risks. Ann N Y Acad Sci 895:245–272
141. Kauffman JM (2003) Radiation hormesis: demonstrated, deconstructed, denied, dismissed, and some implications for public policy. J Scientific Exploration 17:389–407
142. Barros-Dios JM, MS Barreiro, A Ruano-Ravina et al (2002) Exposure to residential radon and lung cancer in Spain: a population-based case-control study. Am J Epidemiol 156: 548–555
143. Lagarde F, G Axelsson, L Damber et al (2001) Residential radon and lung cancer among never-smokers in Sweden. Epidemiology 12:396–404
144. Wichmann H-E, A Schaffrath Rosario, IM Heid et al (2005) Increased lung cancer risk due to residential radon in a pooled and extended analysis of studies in Germany. Health Phys 88:71–79
145. Alavanja MCR, JH Lubin, JA Mahaffey et al (1999) Residential radon exposure and risk of lung cancer in Missouri. Am J Public Health 89:1042–1048
146. Tracy BL, D Krewski, J Chen et al (2006) Assessment and management of residential radon health risks: a report from the Health Canada radon workshop. J Toxicol Environ Health Part A 69:735–758
147. Yarmoshenko I, I Kirdin, M Zhukovsky (2006) Uncertainty analysis of relative biological effectiveness of alpha-radiation for human lung exposure. J Toxicol Environ Health Part A 69:665–679
148. Saccomano G, C Yale, W Dixon et al (1986) An epidemiological analysis of the relationship between exposure to Rn progeny, smoking and bronchogenic carcinoma in the U-mining population of the Colorado plateau-1960–1980. Health Phys 50:605–618
149. Leuraud K, S Billon, D. Bergot et al (2007) Lung cancer risk associated to exposure to radon and smoking in a case-control study of French uranium miners. Health Phys 92:371–378
150. Duport P (2002) Is the radon risk overestimated? Neglected doses in the estimation of the risk of lung cancer in uranium underground miners. Radiat Prot Dosim 98:329–338
151. Saccomanno G, O Auerbach, M Kuschner et al (1996) A comparison between the localization of lung tumors in uranium miners and in non-miners from 1947 to 1991. Cancer 77:1278–1283
152. U.S. Environmental Protection Agency (EPA) (2004) Air toxics risk assessment reference library. vol 1, Technical Resource Manual, Office of Air Quality Planning Standards, Emissions Standards Division.

153. International Agency for Research on Cancer (1997) IARC monographs on the evaluation of carcinogenic risks to humans, vol 68: Silica, some silicates, coal dust and paraaramin fibrils. IARC, Lyon
154. Gilliland FD, WC Hunt, VE Archer et al (2000) Radon progeny exposure and lung cancer risk among non-smoking uranium miners. Health Phys 79:365–372
155 Roscoe RJ, K Steenland, WE Halperin (1989) Lung cancer mortality among nonsmoking uranium miners exposed to radon daughters. J Am Med Assoc 262:629–633
156. Woodward A, D Roder, AJ McMichael et al (1991) Radon daughter exposures at the Radium Hill uranium mine and lung cancer rates among former workers, 1952–87. Cancer Causes Control 2:213–220
157 Tomasek L, A Rogel, M Tirmarche et al (2008) Lung cancer in French and Czech uranium miners: Radon-associated risk at low exposure rates and modifying effects of time since exposure and age at exposure. Radiat Res 169:125–137
158. Xiang-Zhen X, JH Lubin, L Jun-Yao et al (1993) A cohort study in southern China of tin miners exposed to radon and radon decay products. Health Phys 64:120–131
159. Darby SC, E Whitley, GR Howe et al (1995) Radon and cancers other than lung cancer in underground miners: a collaborative analysis of 11 studies. J Natl Cancer Inst 87:378–384
160 Rowlands RE, AF Stheney, HF Lucas (1983) Dose-response relationships for radium-induced bone sarcomas. Health Phys 44(S1):15–31
161. Andersson M, HH Strom (1992) Cancer incidence among Danish thorotrast patients. J Natl Cancer Inst 4:1318–1325
162. Van Kaick G, H Wesch, H Luhrs et al (1991) Neoplastic diseases induced by chronic alpha irradiation. Epidemiological, biophysical and clinical results by the German Thoratrast study. J Radiat Res 32:20–33
163. Tokarskaya ZB, GV Zhuntova, BR Scott et al (2006) Influence of alpha and gamma radiations and non-radiation risk factors on the incidence of malignant liver tumors among Mayak workers. Health Phys 91:296–310
164. Wiggs LD, CA Cox-DeVore, G Voelz (1991) Mortality among a cohort of workers monitored for 210Po exposure: 1944–1972. Health Phys 61:71–76
165. McGeoghegan D, K Binks (2000) The mortality and cancer morbidity experience of workers at the Springfields uranium production facility, 1946–95. J Radiol Prot 20:111–137
166. McGeoghegan D, K Binks (2000) The mortality and cancer morbidity experience of workers at the Capenhurst uranium enrichment facility 1946–95. J Radiol Prot 20:381–401
167. Brown SC, MF Schonbeck, D McClure et al (2004) Lung cancer and internal lung doses among plutonium workers at the Rocky Flats plant: a case-control study. Am J Epidemiol 160:163–172
168. Wilkinson GS, TGL ietjen, LD Wiggs et al (1987) Mortality among plutonium and other radiation workers at a plutonium weapons facility. Am J Epidemiol 125:231–250
169. Tietjen GL (1987) Plutonium and lung cancer. Health Phys 52:625–628
170. Voelz GL, CS Wilkinson, JF Acquavelle (1983) An update of epidemiologic studies of plutonium workers. Health Phys 44(Suppl 1):493–503
171. Voelz GL, JNP Lawrence, ER Johnson (1997) Fifty years of plutonium exposure to the Manhattan project plutonium workers: an update. Health Phys 73:611–619
172. Tokarskaya ZB, ND Okladnikova, ZD Belyaeva et al (1997) Multifactorial analysis of lung cancer dose-response relationships for workers at the Mayak nuclear enterprise. Health Phys 73:899–905
173. Tokarskaya ZB, BR Scott, GV Zhuntova et al (2002) Interaction of radiation and smoking in lung cancer induction among workers at the Mayak nuclear enterprise. Health Phys 83:833–846
174. Koshurnikova NA, MG Bolotnikova, LA Iyin et al (1998) Lung cancer risk due to exposure to incorporated plutonium. Radiat Res 149:366–371

175. Khokryakov V, S Romanov (1994) Lung cancer in radiochemical industry workers. *The Science of the Total Environment* 142:25–28
176. Belyaeva ZD, SV Osovets, BR Scott et al (2008) Modeling of respiratory system dysfunction among nuclear workers: a preliminary study. Dose-Response 6:319–332
177. Jacob V, P Jacob, R Meckbach et al (2005) Lung cancer in Mayak workers: interaction of smoking and plutonium exposure. Radiat Environ Biophys 44:119–129
178. Tokarskaya EB, ND Okladnikova, ZD Belyaeva et al (1995) The influence of radiation and nonradiation factors on the lung cancer incidence among the workers of the nuclear enterprise "Mayak." Health Phys 69:356–366
179. Khokhriakov VF, SA Romanov (1996) Estimation of the temporal distribution and dose dependency of lung cancer among workers of nuclear fuel reprocessing plant. Health Phys 71:83–85
180. Sanders CL, GE Dagle, WC Cannon et al (1976) Inhalation carcinogenesis of high-fired $^{239}PuO_2$ in rats. Radiat Res 68:340–360
181. Sanders CL, D Lundgren (1995) Pulmonary carcinogenesis in the F344 and Wistar rat following inhalation of $^{239}PuO_2$. Rad Res 144:206–214
182. Mitchel REJ, Gragtmans NJ, Morrison DP (1999) Beta-radiation-induced resistance to MNNG initiation of papilloma but not carcinoma formation in mouse skin. Radiat Res 121:180–186
183. Sakai K, Y Hoshi, T Nomura et al (2003) Suppression of carcinogenic processes in mice by chronic low dose rat gamma-irradiation. Intern J Low Radiation 1:142–146

# Breast Cancer 10

"The LNT is dying.... We must accelerate its death because it is a deleterious and costly model" (Maurice Tubiana).

Breast cancer, which markedly increases with age, is the most common cancer among women in the U.S. About one in twenty American women can expect to develop breast cancer during their lifetime [2]. Less than 25% of breast cancers are found in pre-menopausal women [3]. Risk factors for breast cancer in postmenopausal women include age, age at menarche, parity, lactation, age at menopause, diet and calorie intake, obesity, Li-Fraumeni syndrome, family history of breast cancer, height, history of benign breast disease, radiation exposure, prior oral contraceptive use, hormone replacement therapy, and alcohol intake [4–7]. The epidemiology of breast cancer has been widely studied [8].

Breast cancer risk and radiation dose from populations exposed to occupational, medical, and environmental sources of ionizing radiation were obtained from 67 epidemiological studies (Appendix Table A10.1). A great deal of heterogeneity was evident in the pooled data, probably due to imprecision in determining the role of confounding factors. All studies in which breast doses were <10 mSv showed evidence of radiation hormesis. Studies in which breast doses were from >10 mSv to <100 mSv showed breast cancer RR values that were evenly distributed above and below the RR = 1.0 line. Only at breast doses >100 mSv were breast cancer RR values mostly >1.0 (Fig. 10.1).

There is a significantly increased association of breast cancer risk only at radiation doses >100 mGy [10–16]. Atomic bomb survivors exposed to >4 Gy had a RR for breast cancer of 6.4; survivors exposed to doses between 1 and 3 Gy had a RR of 2–4, while those exposed to 0.10–0.49 and 0.50–0.99 Gy had RR values of 1.3 and 1.4, respectively [10, 17]. X-ray treatments for thymus gland enlargement gave an estimated mean dose to the breast of 690 mSv, with a RR for breast cancer of 3.6 [12]. X-ray treatment for ankylosing spondylitis caused significantly increased breast cancer at a mean breast dose of 500 mSv. [13]. Irradiation of children for skin hemangioma was carried out in Stockholm and Gothenburg, Sweden; breast cancer risk was decreased at doses of 0–1,000 mSv to the breast, but increased at doses of 1,000–4,000 mSv [11, 18].

Other epidemiological studies indicate clear evidence of radiation hormesis for breast cancer. In Yangjiang, China, the mean environmental annual dose was 6.4 mSv; the PROFAC for breast cancer (1979–1995) was 44% [19]. The RR for breast cancer in U.S. radiological technologists dropped from 2.08 for those employed before 1950 to 0.76 for

C. L. L. Sanders, *Radiation Hormesis and the Linear-No-Threshold Assumption*,
DOI: 10.1007/978-3-642-03720-7_10, © Springer Verlag Berlin Heidelberg 2010

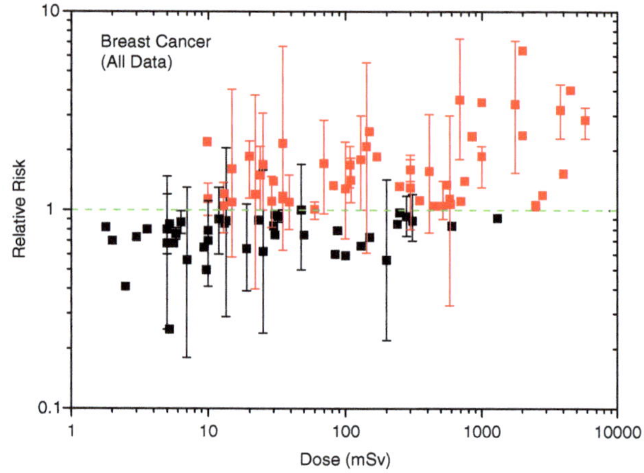


**Fig. 10.1** Relative risk of breast cancer in human populations exposed to ionizing radiation from occupational, medical, and environmental sources

**Table 10.1** Years certified and breast cancer mortality in U.S. radiological technologists [9]

| Year certified | SMR |
|---|---|
| <1940 | 1.5 |
| 1940–49 | 1.1 |
| 1950–59 | 0.9 |
| >1959 | 0.8 |

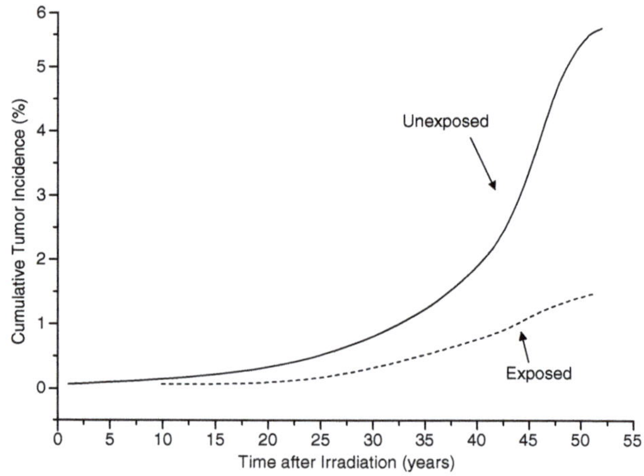


**Fig. 10.2** Cumulative incidence of breast, endometrium, ovary, and prostate cancer in exposed and unexposed children who underwent nasopharyngeal radium therapy for adenoid hypertrophy [21]

those employed from 1970–1979 [20]. A similar trend was seen in another study on U.S. radiological technologists (Table 10.1) [9]. This is an indication of lowered dose limits from early years of employment. Childhood nasopharyngeal radium irradiation for adenoid hypertrophy was associated with cancers of the breast, cervix, uterus, ovary, and prostate, being substantially less than expected (Fig. 10.2) [21, 22]. Pituitary gland irradiation may have altered trophic hormone production, influencing the formation of these tumors.

Miller in a study of breast cancer in TB patients receiving multiple fluoroscopic exams found breast cancer risks substantially <1.0 at breast doses of about 150 and 250 mSv. He

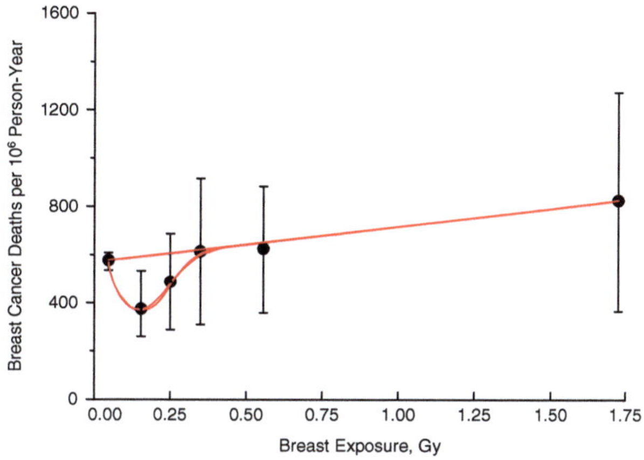


**Fig. 10.3** Breast cancer in tuberculosis patients who received multiple fluoroscopic examinations [23]

**Table 10.2** Relative risk cancer values in white female nuclear weapons workers employed at eleven DOE sites in the U.S. [24]

| Nuclear facility | All cancer mortality | Breast cancer mortality |
|---|---|---|
| Fernald | 0.77 | 0.69 |
| Hanford | 0.78 | 0.85 |
| K-25 | 0.77 | 0.71 |
| Linde | 0.92 | 0.98 |
| Los Alamos | 0.70 | 0.80 |
| Pantex | 0.59 | 0.25 |
| Rocky Flats | 0.60 | 0.68 |
| Savannah River | 0.73 | 0.50 |
| X-10 | 0.75 | 0.82 |
| Y-12 | 0.76 | 0.73 |
| Zia | 0.71 | 0.70 |
| Mean ± S.D. | 0.74 ± 0.09 | 0.70 ± 19.2 |

drew a straight line for the dose-response curve, completely ignoring the two low-dose risk points and then predicted that exposure to 1 cGy at the age of 40 years would increase the number of deaths from breast cancer by 42 per million women. On the contrary, this study clearly demonstrates the beneficial effects of low-dose radiation for breast cancer avoidance and not an increased breast cancer risk associated with the unwarranted use of the LNT assumption (Fig. 10.3) [23].

Wilkinson et al. documented breast cancers in 67,976 women at twelve U.S. nuclear weapons sites. Unbadged women workers were compared with badged women workers. A radiation hormesis response was found in eleven laboratories with a mean PROFAC of 26% for all cancer mortality and 30% for breast cancer mortality. The results could not be explained by the HWE (Table 10.2) [24].

The incidence of breast cancer in the U.S. has more than doubled over the last three decades due, in part, to the increased use of diagnostic mammography in asymptomatic women. Breast cancer screening has reduced breast cancer mortality by about 30% [25]. Diagnostic X-rays

**Table 10.3** Adjusted odds ratio for breast cancer risk in postmenopausal women given one diagnostic series of X-rays [27]

| Diagnostic series | Adjusted odds ratio | 95% CI |
|---|---|---|
| Upper GI series | 0.6 | 0.4–0.8 |
| Lower GI series | 0.8 | 0.5–1.0 |
| Gallbladder series | 0.7 | 0.4–1.0 |
| $^{131}$I Thyroid uptake | 0.9 | 0.6–1.5 |
| Kidney IVP | 0.6 | 0.4–0.9 |
| Cardiac angiogram | 0.9 | 0.6–1.6 |
| Venogram | 0.9 | 0.5–1.8 |
| Myelogram | 1.0 | 0.6–1.8 |
| X-rays broken bones | 0.7 | 0.5–1.0 |

**Table 10.4** Influence of age on breast cancer mortality following mammography screening [34]

| Age cohort | RR | 95% CI |
|---|---|---|
| 40–49 | 0.80 | 0.63–1.01 |
| 50–59 | 0.84 | 0.70–1.01 |
| 60–69 | 0.67 | 0.53–0.84 |
| 65–74 | 0.81 | 0.61–1.07 |

used during the 1940s and 1950s delivered doses that were >50-fold higher than today [26]. A yearly mammogram for the last 30 years resulted in a cumulative radiation dose of ~50 mSv [17]. Many epidemiological studies have been unable to establish low-dose, X-ray exposure as a risk factor for female breast cancer [27]. A case-control study found no evidence of increased breast cancer risk as a result of screening mammography in women at high risk due to carrying the BRCA1 or BRCA2 mutations [28]. In fact, most diagnostic X-ray procedures resulted in less than expected incidences of breast cancer (Table 10.3) [27].

Mammography screening trials in Sweden showed a 24% reduction in breast cancer mortality following breast screening when compared with those who did not receive screening. The reduction in women aged 50–69 years was 29%, but only 13% in women aged 40–49 years [29, 30]. Women aged 40–74 years screened for breast cancer had a 31% reduction in breast cancer mortality and a 25% reduction in the rate of stage II or more advanced breast cancers when compared with those women not screened. The RR for breast cancer mortality was 0.79, while the age-adjusted RR for all cause mortality was 0.98. [29–31].

Is reduction in breast cancer mortality from mammogram screening due to earlier diagnosis (and therefore better prognosis because tumors are diagnosed at an earlier stage) or due to low-dose radiation from annual mammography examinations that suppress breast cancer formation, or both? Screening women has three distinct effects: (1) a large initial increase in rates following initial screening; (2) a persistent, slightly higher risk while women remain in the screening program; and (3) a marked decrease when compared with unscreened women in breast cancer rates when women leave the program [32]. The fall in breast cancer mortality over the last decade has been attributed to both earlier detection and improved treatment [33]. An interesting study of mammographic screening and breast cancer mortality has recently been published, in which the best explanation for the findings was radiation hormesis. The effects of breast cancer screening in terms of breast cancer mortality reduction persist after long-term follow-up (Table 10.4) [34].

## Appendix

**Table A10.5** Risk of breast cancer in epidemiological studies of populations exposed to ionizing radiation

| Dose (mSv) | RR (SMR) | 90–95% CI | Reference |
|---|---|---|---|
| 2.5<br>30<br>100 | 0.41<br>0.83<br>0.59 | –<br>–<br>– | Andersson M, Engholm G, Ennow K et al (1991) Cancer risk among staff at two radiotherapy departments in Denmark. Br J Radiol 64:455–460 |
| 100 | 1.28 | 0.72–2.20 | Blettner M, Zeeb H, Langner I et al (2002) Mortality from cancer and other causes among airline cabin attendants in Germany, 1960–1997. Am J Epidemiol 156:556–565 |
| 150<br>250<br>350<br>450<br>850<br>2,000<br>4,500<br>8,000 | 0.73<br>0.97<br>1.12<br>1.05<br>2.36<br>2.39<br>4.04<br>13.8 | –<br>–<br>–<br>–<br>–<br>–<br>–<br>– | Miller AB, Howe GR, Sherman GJ et al (1989) Mortality from breast cancer after irradiation during fluoroscopic examination in patients being treated for tuberculosis. N Engl J Med 321:1285–1289, (data from all Provinces combined in Canada) |
| 9.3 | 0.65 | – | McGeoghegan D, Binks K (2000) The mortality and cancer morbidity experience of workers at the Springfields uranium production facility, 1946–95. J Radiol Prot 20:111–137 |
| 19 | 0.64 | 0.39–1.07 | Atkinson WD, Law DV, Bromley KJ et al (2004) Mortality of employees of the United Kingdom Atomic Energy Authority, 1946–97. Occup Environ Med 61:577–585 |
| 150 | 2.5 (SIR) | – | Yr W, Sobue T, Lee VS et al (1998) Mortality and cancer incidence in Misasa, Japan, a spa area with elevated radon levels. Jpn J Cancer Res 89:789–796 |
| 32 | 0.93 | 0.87–1.00 | Sont WN, Zielinski JM, Ashmore JP et al (2001) First analysis of cancer incidence and occupational radiation exposure based on the National Dose Registry of Canada. Am J Epidemiol 153:309–318 |
| 551 82 | 1.34<br>1.33 | –<br>– | Wang JX, Zhang LA, Li BX et al (2002). Cancer incidence and risk estimation among medical X-ray workers in China, 1950–1995. Health Phys 82:455–466 |
| 60 | 1.01 | 0.9–1.1 | Mohan AK, Hauptmann M, Freedman DM et al (2003) Cancer and other causes of mortality among radiologic technologists in the United States. Int J Cancer 103:259–267 |

*(continued)*

**Table A10.5** (continued)

| Dose (mSv) | RR (SMR) | 90–95% CI | Reference |
|---|---|---|---|
| 200 | 0.56 | 0.22–1.42 | Tao Z, Zha Y, Akiba S et al (2000) Cancer mortality in the high background radiation areas of Yangjiang, China during the period between 1979 and 1995. J Radiat Res 41(Suppl):31–41 |
| 20 | 1.87 | 1.15–2.23 | Pukkala E, Auvinen A, Wahlberg G (1995). Incidence of cancer among Finnish airline cabin attendants, 1967–92. BMJ 311:649–652 |
| 24 | 1.5 (SIR) | 1.0–2.1 | Rafnsson V, Tulinius H, Jonasson JG, Hrafnkelsson J (2001) Risk of breast cancer in female flight attendants: a population-based study (Iceland). Cancer Causes Control 12:95–101 |
| 39 | 1.1 (SIR) | 0.8–1.5 | Haldorsen T, Reitan JB, Tveten U (2001) Cancer incidence among Norwegian airline cabin attendants. In J Epidemiol 30:825–830 |
| 143 | 2.09 | 0.61–5.55 | Bauer S, Gusev BI, Pivina LM et al (2005) Radiation exposure due to local fallout from Soviet atmospheric nuclear weapons testing in Kazakhstan: solid cancer mortality in the Semipalatinsk historical cohort, 1960–1999. Radiat Res 164:409–419 |
| 413 | 1.57 | 0.77–3.05 | |
| 582 | 1.14 | 0.33–3.00 | |
| 1,769 referent: 20 mSv | 3.43 | 1.54–7.12 | |
| 10 | 0.79 | – | Nyström L, Andersson I, Bjurstam N et al (2002) Long-term effects of mammography screening: updated overview of the Swedish randomised trials. Lancet 359:909–919 |
| 5 | 0.8 | 0.6–1.2 | Zheng T, Holford TR, Mayne ST et al (2002) Radiation exposure from diagnostic and therapeutic treatments and risk of breast cancer. Eur J Cancer Prevent 11:229–236 |
| 12 | 0.9 | 0.6–1.3 | |
| 22 | 1.2 | 0.4–3.8 | |
| 108 | 1.69 | 1.30–2.10 | Morin D, Lonstein M, Stovall JE et al (2000) Breast cancer mortality after diagnostic radiography. Findings from the U.S. scoliosis study. Spine 25:2052–2063 |
| 520 | 1.05 | 0.9–1.4 | Lundell M, Mattsson A, Hakulinen T et al (1996) Breast cancer after radiotherapy for skin hemangioma in infancy. Radiat Res 145:225–230 |
| 170 | 1.86 | – | Lindberg S, Karlsson P, Arvidsson B et al (1995) Cancer incidence after radiotherapy for skin haemangioma during infancy. Acta Oncol 34:735–740 |

**Table A10.5** (continued)

| Dose (mSv) | RR (SMR) | 90–95% CI | Reference |
|---|---|---|---|
| 5,800 | 2.86 | 2.3–3.3 | Mattsson A, Ruden BI, Hall P et al (1993) Radiation-induced breast cancer: Long-term follow-up of radiation therapy for benign breast disease. J Natl Cancer Inst 85:1679–1685 |
| 3,800 | 3.21 | 2.3–4.3 (90% CI) | Shore RE, Hildreth N, Woodward E et al (1986) Breast cancer among women given X-ray therapy for acute postpartum mastitis. J Natl Cancer Inst 77:689–696 |
| 1,000<br>700 | 1.87<br>1.11 | 1.3–2.1<br>– | Boice JD, Preston D, Davis FG et al (1991) Frequent chest X-ray fluoroscopy and breast cancer incidence among tuberculosis patients in Massachusetts. Radiat Res 125:214–222 |
| 750 | 1.40 | – | Davis FG, Boice JD, Hrubec Z, Monson RR (1989) Cancer mortality in a radiation-exposed cohort of Massachusetts tuberculosis patients. Cancer Res 49:6130–6136 |
| 300 | 1.6 | 1.4–1.8 | Thompson DE, Mabuchi K, Ron E et al (1994) Cancer incidence in atomic bomb survivors. Part II: Solid tumors, 1958–1987. Radiat Res 137(Suppl):S17–S67 |
| 690 | 3.60 | 1.8–7.3 | Hildreth NG, Shore RE, Dvoretsky PM (1989) The risk of breast cancer after irradiation of the thymus in infancy. NEJM 321:1281–1284 |
| 2,510 | 1.04 | – | Storm HH, Andersson M, Boice JD et al (1992) Adjuvant radiotherapy and risk of contralateral breast cancer. J Natl Cancer Inst 84:1245–1250 |
| 2,820 | 1.19 | – | Boice JD, Harvey EB, Blettner M et al (1992) Cancer in the contralateral breast after radiotherapy for breast cancer. N Engl J Med 326:781–785 |
| 250 | 1.32 | – | Pierce DA, Shimizu Y, Preston DL et al (1996) Studies of the mortality of atomic bomb survivors. Report 12, Part I. Cancer: 1950–1990. Radiat Res 146:1–27 |
| 590 | 1.07 | – | Weiss HA, Darby SC, Doll R (1994) Cancer mortality following x-ray treatment for ankylosing spondylitis. Int J Cancer 59:327–338 |
| 1,000 (RBE = 10) | 3.5 | – | Nekolla EA, Kellerer AM, Kuse-Isingschulte M et al (1999) Malignancies in patients treated with high doses of radium-224. Radiat Res 152:3–7 |

(*continued*)

**Table A10.5** (continued)

| Dose (mSv) | RR (SMR) | 90–95% CI | Reference |
|---|---|---|---|
| 5 | 0.68 | – | Møller B, Weedon-Fekjær H, Hakulinen T et al (2005) The influence of mammographic screening on national trends in breast cancer incidence. Eur J Cancer Prevent 14:117–128 |
| 50 | 0.75 | – | Mifune M, Sobue T, Arimoto H, et al (1992) Cancer mortality survey in a spa area (Misasa, Japan) with a high radon background. Jpn J Cancer Res 83:1–5 |
| 600 | 0.83 | – | Doody MM, Mandel JS, Lubin JH et al (1998). Mortality among United States radiologic technicians, 1926–90. Cancer Causes Control 9:67–75 |
| 1,300 | 0.91 | – | |
| 2,500 | 1.06 | – | |
| 4,000 | 1.53 | – | |
| 13 | 1.05 | 0.77–1.38 | Schubauer-Berigan MK, Macievic GV, Utterback DF et al (2005) Final Report. An epidemiologic study of mortality and radiation-related risk of cancer among workers at the Idaho National Engineering and Environmental Laboratory, a U.S. Department of Energy facility. NIOSH |
| 5.8 | 0.76 | 0.71–0.81 | Wilkinson GS, Trieff N, Graham R et al (2000) Final Report. Study of mortality among female nuclear weapons workers. Grant Numbers: 1R01 OHO3274, R01/CCR2144546, R01/CCR61 2934–01, National Institute for Occupational Safety and Health, Centers for Disease Control and Prevention |
| 0.65 | 0.69 | – | |
| 5.2 | 0.85 | – | |
| 0.82 | 0.71 | – | |
| 3.6 | 0.80 | – | |
| 5.2 | 0.25 | – | |
| 5.6 | 0.68 | – | |
| 9.7 | 0.50 | – | |
| 1.8 | 0.82 | – | |
| 3.0 | 0.73 | – | |
| 2.0 | 0.70 | – | |
| 13.5 | 0.88 | 0.29–2.05 | Boice JD, Cohen SS, Mumma MT et al (2006) Mortality among radiation workers at Rocketdyne (Atomics International), 1948–1999. Radiat Res 166:98–115 |
| 25 | 0.62 | 0.24–1.59 | Yeh H, Matanoski GM, Wang N et al (2001) Cancer incidence after childhood nasopharyngeal irradiation: a follow-up study in Washington County, Maryland. Am J Epidemiol 153:749–756 |
| 5 | 0.68 | 0.25–1.48 | Ahrenholz S, Cardarelli J, Dill P et al (2001) Final Report. Mortality patterns among workers at the Portsmouth gaseous diffusion plant Piketon, Ohio. NIOSH |
| 109 | 0.0 | 0.00–18.59 | Rahu M, Rahu K, Auvinen A et al (2006) Cancer risk among Chernobyl cleanup workers in Estonia and Latvia, 1986–1998. Int J Cancer 119:162–168 |

**Table A10.5** (continued)

| Dose (mSv) | RR (SMR) | 90–95% CI | Reference |
|---|---|---|---|
| 7 | 0.56 | 0.18–1.30 | Silver SR, Anderson-Mahoney P, Burphy J et al (2005) Mortality update for the Pantex weapons facility: final report health-related energy research branch. National Institute for Occupational Safety and Health, HHS (NIOSH) Publication No. 2005–124 |
| 23.3 | 0.89 | – | Gilbert ES, Omohundro E, Buchanan JA et al (1993) Mortality of workers at the Hanford site: 1945–1986. Health Phys 64:577–590 |
| 35 | 2.17 | 0.63–6.70 | Carpenter LM, Higgins CD, Douglas AJ et al (1998) Cancer mortality in relation to monitoring for radionuclide exposure in three UK nuclear industry workforces. Br J Cancer 78:1224–1232 |
| 10 | 0.70 (SIR) | 0.41–1.11 | Ron E, Auvinen A, Alfandary E et al (1999) Cancer risk following radiotherapy for infertility or menstrual disorders. Int J Cancer 82:795–798 |
| 25 | 1.7 (SIR) | 1.0–3.1 | Jartti P, Pukkala E, Uitti J et al (2006). Cancer incidence among physicians occupationally exposed to ionizing radiation in Finland. Scand J Work Environ Health 32:368–373 |
| 280 | 0.93 | 0.74–1.18 | Franklyn JA, Maisonneuve P, Sheppard M et al (1999) Cancer incidence and mortality after radioiodine treatment for hyperthyroidism: a population-based cohort study. Lancet 353:2111–2115 |
| 300 | 1.3 | 0.8–1.9 | Goldman MB, Maloof F, Monson RR et al (1988) Radioactive iodine therapy and breast cancer. Am J Epidemiol 127:969–980 |
| 35 | 1.16 | 1.09–1.23 | Sigurdson AJ, Doody MM, Rao RS et al (2003) Cancer incidence in the U.S. radiologic technologists health study, 1983–1998. Cancer 97:3080–3089 |
| 110 | 1.42 | 1.09–1.83 | Reynolds P, Cone J, Layefsky M et al (2002) Cancer incidence in California flight attendants (United States). Cancer Causes Control 13:317–324 |
| ~10 | 1.14 | 0.94–1.37 | Telle-Lamberton M, Bergot D, Gagneau M et al (2004) Cancer mortality among French Atomic Energy Commission workers. Am J Indust Med 45:34–44 |

*(continued)*

**Table A10.5** (continued)

| Dose (mSv) | RR (SMR) | 90–95% CI | Reference |
|---|---|---|---|
| 6.3 | 0.87 | 0.76–0.99 | Ashmore JP, Krewski D, Ziellnski JM et al (1998) First analysis of mortality and occupational radiation exposure based on the National Dose Registry of Canada. Am J Epidemiol 148:564–574 |
| 13.2 | 0.86 | – | Frome EL, Cragle DL, Watkins JP et al (1997) A mortality study of employees of the nuclear industry in Oak Ridge, Tennessee. Radiat Res 148:64–80 |
| 15 | 1.10 | – | Beral V, Fraser P, Carpenter L et al (1988) Mortality of employees of the Atomic Weapons establishment, 1951–1982. BMJ 297:757–770 |
| 30.5 | 0.75 | – | Muirhead CR, Goodill AA, Haylock RGE et al (1999) Occupational radiation exposure and mortality: second analysis of the National Registry of Radiation Workers. J Radiol Prot 19:3–26 |
| 310 | 0.88 | 0.7–1.2 | Boice JD, Blettner M, Kleinerman RA et al (1989) Radiation dose and breast cancer risk in patients treated for cancer of the cervix. Int J Cancer 44:7–16 |
| 87.4 | 0.79 | – | Omar RZ, Barber JA, Smith PG (1999) Cancer mortality and morbidity among plutonium workers at the Sellafield plant of British Nuclear Fuels. Br J Cancer 79:1288–1301 |
| 30.7 | 0.95 | – | Fraser P, Carpenter L, Maconochie N et al (1993) Cancer mortality and morbidity in employees of the United Kingdom Atomic Energy Authority, 1946–1986. Br J Cancer 67:615–624 |
| 13.2 | 1.21 | – | Loomis DP, Wolf SH (1996) Mortality of workers at a nuclear materials production plant at Oak Ridge, Tennessee, 1947–1990. Am J Ind Med 29:131–141 |
| 47.8 | 1.0 (SIR) | 0.5–1.7 | Hwang SL, Guo HR, Hsieh WA et al (2006) Cancer risks in a population with prolonged low-dose-rate γ-radiation exposure in radiocontaminated buildings, 1983–2022. Int J Radiat Biol 82:849–858 |
| 29 | 1.11 | 0.82–1.48 | Zeeb H, Blettner M, Langner I et al (2003) Mortality from cancer and other causes among airline cabin attendants in Europe: a collaborative cohort study in eight countries. Am J Epidemiol 158:35–46 |
| 30 | 1.40 | – | Lynge E (2001) Commentary: cancer in the air. Int J Epidemiol 30:830–832 |

**Table A10.5** (continued)

| Dose (mSv) | RR (SMR) | 90–95% CI | Reference |
|---|---|---|---|
| 15 | 1.61 (SIR) | 0.58–4.08 | Habib RR, Abdallah SM, Law M, Kaldor J (2006) Cancer incidence among Australian nuclear industry workers. J Occup Health 48:358–365 |
| 70 | 1.72 (SIR) | 0.96–2.84 | Ivanov VK, Tsyb AF, Rastopchin EM et al (2001) Cancer incidence among nuclear workers in Russia based on data from the Institute of Physics and Power Engineering: a preliminary analysis. Radiat Res 155:801–808 |
| 84 | 0.60 | – | McGeoghegan D, Binks K (2001) The mortality and cancer morbidity of employees at the Chapelcross plant of British Nuclear Fuels plc, 1955–95. J Radiol Prot 21:221–250 |
| 9.9 | 2.21 | – | McGeoghegan D, Binks K (2000) The mortality and cancer mortality experience of workers at the Capenhurst uranium enrichment facility 1946–95. J Radiol Prot 20:381–401 |
| ~3,000 | 75.3 | 44.9–118.4 | Bhatia S, Robison LL, Oberlin O et al (1996) Breast cancer and other second neoplasms after childhood Hodgkin's Disease. New Engl J Med 334:745–751 |
| 300 | 1.3 | – | John MJ, Kelsey JL (1993) Radiation and other environmental exposures and breast cancer. Epidemiol Rev 15:157–162 |
| 750 | 1.4 | – | |
| 2,000 | 6.4 | – | |
| 130 | 1.8 (SIR) | 1.0–3.0 | Hoffman DA, Lonstein JE, Morin MM et al (1989) Breast cancer in women with silicosis to multiple diagnostic X-rays. J Natl Cancer Inst 81:1312–1389 |

## References


1. Zahl P-H, Maehlen J, Welch G (2008) The natural history of invasive breast cancers detected by screening mammography. Arch Intern Med 168:2311–2316
2. Boring CC, Squires TS, Tong T, Montgomerie S (1994) Cancer statistics 1994. CA Cancer J Clin 44:7–26
3. Marchant DJ (1994) Supplemental estrogen replacement. Cancer 74:512–517
4. Key TJ, Pike MC (1988) The role of estrogens and progestagens in the epidemiology and prevention of breast cancer. Eur J Cancer Clin Oncol 24:29–43
5. Harris JR, Lippman ME, Veronesi U, Willett W (1992) Breast cancer. N Engl J Med 327:319–328

6. de Waard F, Trichopoulos D (1988) A unifying concept of the aetiology of breast cancer. Int J Cancer 41:666–669
7. Tryggvadottir L, Tulinius H, Eyfjord JE et al (2002) Breast cancer risk factors and age at diagnosis: an Icelandic cohort study. Int J Cancer 98:604–608
8. Kelsey JL (1993) Breast cancer, epidemiologic reviews, vol 15. Johns Hopkins University School of Hygiene and Public Health, Baltimore, MD
9. Doody MM, Mandel JS, Lubin JH et al (1998) Mortality among United States radiologic technicians, 1926–90. Cancer Causes Control 9:67–75
10. Preston DL, Mattsson A, Holmberg E et al (2002) Radiation effects on breast cancer risk: a pooled analysis of eight cohorts. Radiat Res 158:220–235
11. Lundell M, Mattsson A, Hakulinen T, Holm LE (1996) Breast cancer after radiotherapy for skin hemangioma in infancy. Radiat Res 145:225–230
12. Hildreth NG, Shore RE, Dvoretsky PM (1989) The risk of breast cancer after irradiation of the thymus in infancy. N Engl J Med 321:1281–1284
13. Darby SC, Doll R, Gill SK, Smith PG (1987) Long term mortality after a single treatment course with X-rays in patients treated for ankylosing spondylitis. Br J Cancer 55:180–190
14. Hoffman DA, Lonstein JE, Morin MM et al (1989) Breast cancer in women with scoliosis exposed to multiple diagnostic X-rays. J Natl Cancer Inst 81:1312–1389
15. Keirim-Markus IB (2002) Radiation exposure normalization taking account of specific effects at low doses and dose rates. Atomic Energy 93:836–844
16. Ronckers CM, Doody MM, Lonstein JE et al (2008) Multiple diagnostic X-rays for spine deformities and risk of breast cancer. Cancer Epidemiol Biomarkers Prev 17:605–613
17. John MJ, Kelsey JL (1993) Radiation and other environmental exposures and breast cancer. Epidemiol Rev 15:157–162
18. Lundell M, Mattson A, Holmberg P et al (1999) Breast cancer risk after radiotherapy in infancy: a pooled analysis of two Swedish cohorts of 17,202 infants. Radiat Res 151:626
19. Tao Z, Zha Y, Akiba S et al (2000) Cancer mortality in the high background radiation area of Yangjiang, China during the period between 1979 and 1995. J Radiat Res 41(Suppl): 31–41
20. Mohan AK, Hauptmann M, Freedman DM et al (2003) Cancer and other causes of mortality among radiologic technologists in the United States. Int J Cancer 103:259–267
21. Yeh H-C, Matanoski GM, Wang N-Y et al (2001) Cancer incidence after childhood nasopharyngeal radium irradiation: a follow-up study in Washington County, Maryland. Am J Epidemiol 153:749–756
22. Ronckers CM, van Leeuwen FE, Hayes RB et al (2002) Cancer incidence after nasopharyngeal radium irradiation. Epidemiology 13:552–560
23. Miller AB, Howe GR, Sherman GJ et al (1989) Mortality from breast cancer after irradiation during fluoroscopic examination in patients being treated for tuberculosis. N Engl J Med 321:1285–1289
24. Wilkinson GS, Trieff N, Graham R et al (2000) Final Report. Study of mortality among female nuclear weapons workers. Grant Numbers: 1R01 OHO3274, R01/CCR2144546, R01/CCR61 2934-01, National Institute for Occupational Safety and Health, Centers for Disease Control and Prevention
25. Berlin NI (1995) Breast cancer screening between ages 40 and 49. Cancer J Sci Am 1:187–190
26. Cardarelli JJ, Spitz HB, Rice C et al (2002) Evaluation of work-related medical X rays in epidemiological studies of nuclear workers. Radiat Res 158:807–808
27. Zheng T, Holford S, Mayne T et al (2002) Radiation exposure from diagnostic and therapeutic treatments and risk of breast cancer. Eur J Cancer Prev 11:229–235
28. Narod SA, Lubinski J, Ghadrinian P et al (2006) Screening mammography and the risk of breast cancer in BRCA1 and BRCA2 mutation carriers: a case-control study. Lancet Oncol 7:402–406

29. Tabar L, Vitak B, Chen H-H et al (2000) The Swedish two-county trial twenty years later: updated mortality results and new insights from long-term follow-up. Radiol Clin North Am 38:625–651
30. Nystrom L, Wall S, Rutqvist LE et al (1993) Breast cancer screening with mammography: overview of Swedish randomised trials. Lancet 341(8851):973–978
31. Tabar L, Gad A, Holmberg LH et al (1985) Reduction in mortality from breast cancer screening with mammography. Randomised trial from the Breast Cancer Screening Working Group of the Swedish National Board of Health and Welfare. Lancet 325(8433):829–832
32. Moller B, Weedon-Fekjaer H, Hakulinen T et al (2005) The influence of mammographic screening on national trends in breast cancer incidence. Eur J Cancer Prev 14:117–128
33. Levi F, Bosetti C, Lucchini F et al (2005) Monitoring the decrease in breast cancer mortality in Europe. Eur J Cancer Prev 14:497–502
34. Nystrom L, Andersson I, Bjurstam N et al (2002) Long-term effects of mammography screening: updated overview of the Swedish randomized trials. Lancet 359:909–919

# Leukemia 11

> No one has been identifiably injured by radiation while working within the first numerical standards set first by the NCRP and then the ICRP in 1934
>
> (Lauriston Taylor)

Leukemia is a disease of abnormal lymphopoiesis, myelopoiesis, or erythropoiesis resulting in multicentric and unexplained proliferation and/or accumulation of neoplastic cells. The basic problem with leukemia is one of departure from steady-state equilibrium in the production of marrow cells and the resultant infiltration and encroachment of excess leukemia cells upon other tissues, interfering with their normal function.

The risk of leukemia is associated with age, sex, geographic location, genetics and exposure to bracken fern, benzene and other chemicals, and ionizing radiation. The risk of developing leukemia is 1 in 5 for an identical twin if the other has leukemia, 1 in 6 for those treated with $^{32}P$ for polycythemia vera, 1 in 8 for those with Bloom's syndrome, 1 in 60 for Hiroshima A-bomb survivors within 1,000 m of hypocenter, 1 in 95 for those with Down's syndrome, 1 in 720 for siblings of leukemia children, and 1 in 2,880 for children <15 years old [1, 2]. Leukemia is also a late complication of combined cancer radiotherapy and chemotherapy [3, 4].

There is no convincing evidence that populations living in high background radiation regions, around nuclear reactors, in contaminated regions around nuclear accident sites (Eastern Urals, Chernobyl), and near nuclear weapon test sites experience an increase in leukemia [5]. There is little evidence of increased leukemia at doses <100 mSv in most epidemiological studies of populations exposed to a variety of radiation sources. Among these populations are residents of high-dose backgrounds, tuberculosis patients given multiple fluoroscopies, women given annual mammograms, Japanese A-bomb survivors, Chernobyl emergency workers, Semipalatinsky nuclear test-site participants, Mayak nuclear workers, downwind inhabitants of the Eastern Urals nuclear waste tank explosion, Techa River inhabitants, radiologists and radiological technicians, children and military personnel given nasopharyngeal radium treatment for inflamed adenoids and tonsils, patients receiving diagnostic and therapeutic doses of $^{131}I$, radium dial painters, thorotrast patients, German radium therapy patients, uranium and hard rock miners, and from residential radon [6].

A large heterogeneity in risk of leukemia mortality was found in a pooled analysis of 75 published studies of populations exposed occupationally, medically, accidentally, or

C. L. L. Sanders, *Radiation Hormesis and the Linear-No-Threshold Assumption*,
DOI: 10.1007/978-3-642-03720-7_11, 

environmentally to ionizing radiation. Leukemia mortality risk was equally distributed above and below the RR = 1.0 line for dose bins <100 mSv. Only for dose bins >100 mSv, a predominance in increased leukemia mortality (Appendix Table A11.3; Fig. 11.1) was observed. The pooled study shows neither a net risk nor benefit from cumulative exposures <100 mSv for all exposed populations or for only nuclear workers (Figs. 11.1 and 11.2).

Individual studies, however, do show evidence of radiation hormesis with less than expected leukemia mortality at cumulative doses <100 mSv from low LET radiation exposure. The mean dose for participants in OPERATION GREENHOUSE, a U.S. nuclear test, was 13 mSv and the SMR for leukemia was 0.23 [7]. Less than expected leukemia and non-Hodgkin's lymphoma were observed in children whose fathers received doses <100 mSv (Table 11.1) [5, 8]. Radiation exposure may have induced apoptosis of genomically damaged spermatogonia. Leukemia was not increased in children whose fathers were Japanese A-bomb survivors exposed prior to conception [8].

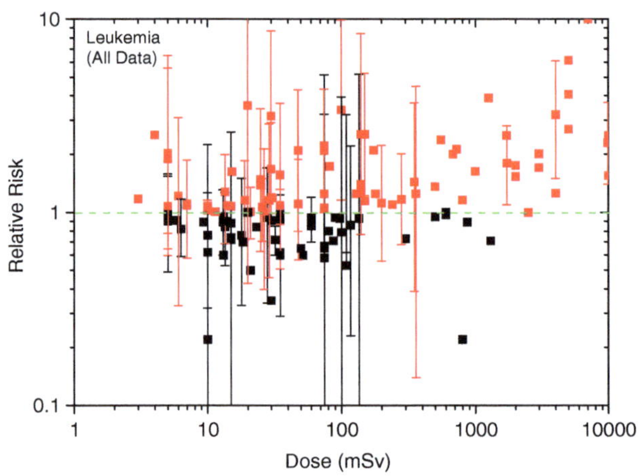


**Fig. 11.1** Relative risk of leukemia mortality from all sources of radiation exposure

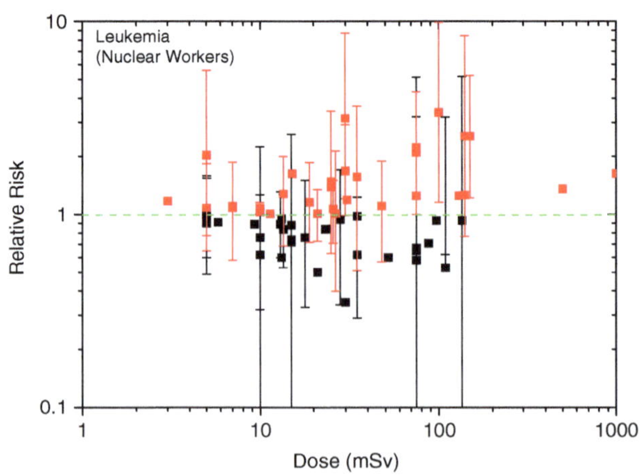


**Fig. 11.2** Relative risk of leukemia mortality only in nuclear workers

**Table 11.1** Relative risk of childhood leukemia and non-Hodgkin's lymphoma associated with preconception paternal irradiation [8]

| Population | Relative risk | 95% CI |
|---|---|---|
| Sellafield | 0.73 | 0.47–2.08 |
| Scotland | 0.51 | 0.51–2.95 |
| Ontario | 0.63 | 0.27–3.40 |

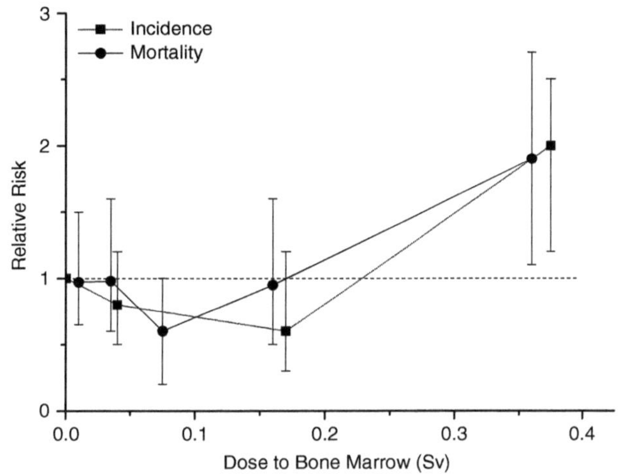


**Fig. 11.3** Relative risk of leukemia in Japanese A-bomb survivors [18, 19]

The only study where data was in favor of the LNT assumption was the Oxford study regarding childhood cancer and *in utero* exposure [9]. However, this study has been criticized and the results refuted in five subsequent studies [10–14], including a large case–control study [11] and the findings of an IRAC committee [14].

Leukemia was the first malignancy associated with radiation exposure in A-bomb survivors [15, 16]. Leukemia mortality increased fivefold at doses >1,000 mSv and 14-fold at doses >2,000 mSv. The dose cohort of >1,000 mSv represents only 3.2% of the A-bomb survivor population. The cohort of 5–100 mSv represents 44.5% of the survivor population; leukemia mortality in this group was decreased by about 50%. A table in UNSCEAR (1958) showed that leukemia incidence in Hiroshima survivors was decreased by 66% in those exposed to 20 mSv compared with the unexposed controls [17]. Even so, the BEIR VII committee using the LNT assumption and high-dose data, estimated that there would be an excess of 170 leukemia cases from exposure of A-bomb survivors to 100 mSv [5]. The clear presence of a threshold and radiation hormesis for leukemia in Japanese A-bomb survivors at exposures of <200 mSv was ignored by BEIR VII (Fig. 11.3) [18, 20].

A significant negative association between acute myeloid leukemia mortality in adults and $\gamma$-radiation were found in 41 of 95 French "Departements" [21]. A negative nonsignificant association between childhood leukemia incidence and $\gamma$-radiation was observed in 22 counties of the UK [22]. The incidence of leukemia and lymphoma were 19% less in males and 6% less in females living in the U.S. at an altitude of 2,000–5,300 feet when compared with those living at an altitude of <500 feet [23].

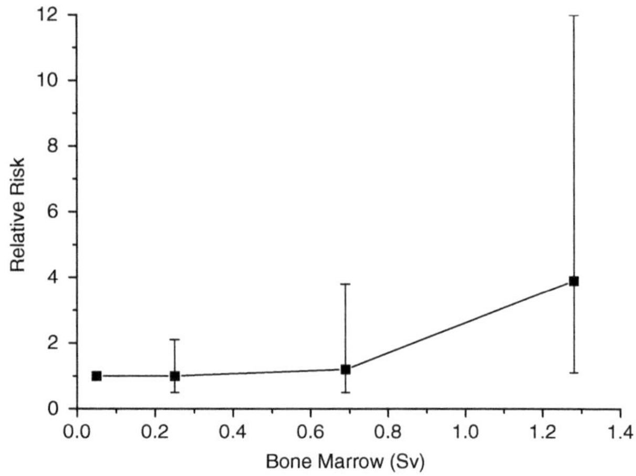


**Fig. 11.4** Relative risk for leukemia among residents along the Techa River, excluding chronic lymphoid leukemia [32]

The lack of a positive association between childhood leukemia and indoor radon are summarized in several publications [24–26]. A meta-analysis of five published case–control studies found no evidence of an association between residential radon and childhood leukemia [27]. A nonsignificant increase was found for leukemia in studies of uranium miners [28, 29]. Other studies have shown a large decrease in leukemia among miners. A uranium miner study in China, where the mean lung dose was 678 mSv, showed a PROFAC of 67% for leukemia [30]. A study of radon in Brazilian coal miners showed a PROFAC of 61% for leukemia at a mean lung dose of 140 mSv [31].

About 30,000 people lived along the banks of the Techa River during the early years (1949–1956) nearby the Mayak nuclear weapons facility in the Southern Urals. Radiation exposures involved both external and internal sources from contaminated river water and ingested food. Significant leukemia risk occurred only for doses >0.5 Gy to bone marrow (Fig. 11.4) [33]. Increased leukemia incidence was found only in Mayak workers exposed to cumulative doses >3 Sv over a 5-year period of time [34].

Leukemia is rarely associated with exposure to plutonium and other transuranics [35]. $^{224}$Ra was used in the therapy of patients with bone tuberculosis and spondylitis. Colloidal $^{232}ThO_2$ (Thorotrast) was used as an X-ray contrast medium up until 1955. Leukemia in radium dial painters, thorotrast patients, and radium therapy patients was seen only at high bone doses [36]. No leukemia deaths were found in female British dial painters. The British radium dial painters, like the U.S. dial painters, exhibited an increased lifespan [37]. Radium dial painters received an average bone marrow dose of 40 mSv/year. The PROFAC for leukemia in 1,285 US dial painters was 78% [38].

A significantly increased incidence of leukemia was not found among Chernobyl clean-up workers [39, 40]. In fact, the overall mortality rate was significantly lower for clean-up staff than for the general public [41]. The incidence of childhood or adulthood leukemia was not increased as a result of exposures from the Chernobyl accident [5, 42]. Among the dose bins, ranging from 17 to 215 mSv for Chernobyl emergency workers, there is evidence for both radiation hormesis and increased risk of leukemia [43] (Table 11.2).

**Table 11.2** Risk of leukemia in Chernobyl emergency workers (95% CI) [43]

| Mean Dose (mGy) | 1986–1996 Cohort | 1997–2003 Cohort |
|---|---|---|
| 17 | 1.0 | 1.0 |
| 66 | 0.4 (0.1–1.0) | 1.1 (0.5–2.6) |
| 106 | 0.4 (0.1–1.0) | 0.6 (0.2–1.5) |
| 215 | 1.4 (0.8–2.6) | 0.9 (0.3–1.8) |

Mortality was examined in workers employed in 15 nuclear power utilities in the U.S. between 1979 and 1997; no significant associations were seen for leukemia[44]. In a cohort of 45,468 Canadian nuclear power industry workers (1957–1994), the PROFAC for leukemia was 32% [45]. No association between cumulative radiation dose and leukemia mortality was found in 50,000 UKAEA workers [46]. In a combined study of nuclear workers from the U.S., UK, and Canada, it was claimed that leukemia was significantly associated with cumulative external radiation dose. This claim was based on only eight cases of leukemia spread among four dose groups. A decreased leukemia incidence was seen at a dose range of 20–40 mSv (PROFAC = 27%) and an increased risk only at doses >40 mSv [47]. No association between external radiation exposure and chronic lymphocytic leukemia mortality was observed in the 15-country study of nuclear industry workers [48].

## Appendix

**Table A11.3** Risk of leukemia mortality in epidemiological studies of populations exposed to ionizing radiation

| Dose, mSv | RR (SMR) | 90–95% CI | References |
|---|---|---|---|
| 5 | 1.9 | 0.6–6.5 | Ronckers CM, van Leewen FE, Hayes RB et al (2002) Cancer incidence after nasopharyngeal radium irradiation. Epidemiol 13:552–560 |
| 19 | 1.16 | 0.72–1.86 | Atkinson WD, Law DV, Bromley KJ et al (2004) Mortality of employees of the United Kingdom Atomic Energy Authority, 1946–97. Occup Environ Med 61:577–585 |
| 5 | 0.90 | – | |
| 15 | 0.88 | – | |
| 35 | 0.98 | – | |
| 75 | 1.25 | – | |
| 130 | 1.25 | – | |
| 240 | 1.1 | – | Ostroumova E, Gagniere B, Lauier D et al (2006) Risk analysis of leukaemia incidence among people living along the Techa River: a nested case–control study. J Radiol Prot 26:17–32 |
| 680 | 2.0 | – | |
| 1,260 | 3.9 | – | |
| 4,000 | 1.26 | – | Doody MM, Mandel JS, Lubin JH et al (1998) Mortality among United States radiologic technicians, 1926–90. Cancer Causes Control 9:67–75 |
| 2,500 | 1.00 | – | |
| 1,300 | 0.71 | – | |
| 600 | 0.97 | – | |

(*continued*)

**Table A11.3** (continued)

| Dose, mSv | RR (SMR) | 90–95% CI | References |
|---|---|---|---|
| 551 | 2.37 | – | Wang JX, Zhang LA, Li BX et al (2002) Cancer incidence and risk estimation among medical X-ray workers in China, 1950–1995. Health Phys 82:455–466 |
| 82 | 1.73 | – | |
| 200 | 1.12 | 0.56–2.22 | Tao Z, Zha Y, Akiba S et al (2000) Cancer mortality in the high background radiation areas of Yangjiang, China during the period between 1979 and 1995. J Radiat Res 41(suppl):31–41 |
| 10 | 1.03 | – | McGeoghegan D, Binks K (2000) The mortality and cancer morbidity experience of workers at the Capenhurst uranium enrichment facility 1946–95. J Radiol Prot 20:381–401 |
| 5 | 0.95 | – | Cardis E, Gilbert ES, Carpenter L et al (1995) Effects of low doses and low dose rates of external ionizing radiation: cancer mortality among nuclear industry workers in three countries. Radiat Res 142:117–132 |
| 15 | 1.08 | – | |
| 35 | 0.92 | – | |
| 75 | 1.06 | – | |
| 150 | 1.15 | – | |
| 300 | 0.73 | – | |
| 26.5 | 1.01 | 0.40–2.13 | Wiggs LD, Cox-DeVore CA, Voelz GL (1991) Mortality among a cohort of workers monitored for 210Po exposure: 1944–1972. Health Phys 61:71–76 |
| 25.7 | 1.07 | 0.71–1.5 | Howe GR, Zablotska LB, Fix JJ et al (2004) Analysis of the mortality experience amongst U.S. nuclear power industry workers after chronic low-dose exposure to ionizing radiation. Radiat Res 162:517–526 |
| 25 | 1.48 | 0.63–3.44 | |
| 75 | 0.65 | 0.08–5.16 | |
| 140 | 2.54 | 0.77–8.44 | |
| 13.5 | 0.84 | 0.53–1.27 | Zablotska LB, Ashmore JP, Howe GR (2004) Analysis of mortality among Canadian nuclear power industry workers after chronic low-dose exposure to ionizing radiation. Radiat Res 161:633–641 |
| 97 | 0.93 | – | Sponsler R, Cameron JR (2005) Nuclear shipyard worker study (1980–1988): a large cohort exposed to low-dose-rate gamma radiation. Intern J Low Radiat 1:463–478 |
| 32 | 0.72 (SIR) | 0.60–0.85 | Sont WN, Zielinski JM, Ashmore JP et al (2001) First analysis of cancer incidence and occupational radiation exposure based on the National Dose Registry of Canada. Am J Epidemiol 153:309–318 |

**Table A11.3** (continued)

| Dose, mSv | RR (SMR) | 90–95% CI | References |
|---|---|---|---|
| 117 | 0.86 | 0.23–2.2 | Band PR, Le ND, Fang R et al (1996) Cohort study of Air Canada pilots: mortality, cancer incidence, and leukemia risk. Am J Epidemiol 143:137–143 |
| 21 | 0.50 | – | Jablon S, Boice JD (1993) Mortality among workers at a nuclear power plant in the United States. Cancer Causes Control 4:427–430 |
| 6 | 1.22 | 0.33–3.1 | Muirhead CR, Kendall GM, Darby SC et al (2004) Epidemiological studies of UK test veterans: II. Mortality and cancer incidence. J Radiol Prot 24:219–241 |
| 140 | 1.40 | – | Ye W, Sobue T, Lee VS et al (1998) Mortality and cancer incidence in Misasa, Japan, a spa area with elevated radon levels. Jpn J Cancer Res 89:789–796 |
| 20 | 3.57 | 0.43–12.9 | Pukkala E, Auvinen A, Wahlberg G (1995) Incidence of cancer among Finnish airline cabin attendants, 1967–92. Br Med J 311:649–652 |
| 109 | 1.53 | 0.62–3.2 | Rahu M, Rahu K, Auvinen A et al (2006) Cancer risk among Chernobyl cleanup workers in Estonia and Latvia, 1986–1998. Int J Cancer 119:162–168 |
| 28 | 0.94 | 0.34–1.7 | Rinsky RA, Zumwalde RD, Waxweiler RJ et al (1981) Cancer mortality at a naval nuclear shipyard. Lancet 317:231–235 |
| 90 | 0.94 | – | Davis FG, Boice J, Hrubec D et al (1989) Lung cancer Massachusetts tuberculosis patients. Cancer Res 49:6130–6136 |
| 14 | 0.90 | – | Epidemiological Study Group of Nuclear Workers (Japan) (1997) First analysis of mortality of nuclear industry workers in Japan, 1986–1992. J Health Phys 32:173–184 |
| 20 | 1.0 | – | UNSCEAR (2000) Sources and effects of ionizing radiation. United Nations Scientific Committee on the effects of atomic radiation. 2000 report to the General Assembly with annexes. Volume II: Effects NO E.OO.IX.4, United Nations, NY. |
| 35 | 0.6 | – | |
| 80 | 0.8 | – | |
| 175 | 2.1 | – | |
| 370 | ? | – | |
| 720 | 2.12 | – | Koshurnikova NA, Bysogolov GD, Bolotnikova MG et al (1996) Mortality among personnel who worked at the MAYAK complex in the first years of operation. Health Phys 71:90–93 |

*(continued)*

**Table A11.3** (continued)

| Dose, mSv | RR (SMR) | 90–95% CI | References |
|---|---|---|---|
| 870 | 0.89 | – | |
| 1,710 | 2.50 | – | |
| 14 | 1.09 | – | Hall P, Boice JD, Berg G et al (1992) Leukaemia incidence after iodine-131 exposure. Lancet 340:1–4 |
| 9.3 | 0.89 | – | McGeoghegan D, Binks K (2000) The mortality and cancer morbidity experience of workers at the Springfields uranium production facility, 1946–95. J Radiol Prot 20:111–137 |
| 100 | 0.79 | 0.06–3.94 | Blettner M, Zeeb H, Langner I et al (2002) Mortality from cancer and other causes among airline cabin attendants in Germany, 1960–1997. Am J Epidemiol 156:556–565 |
| 13 | 0.89 | 0.60–1.31 | Schubauer-Berigan MK, Macievic GV, Utterback DF et al (2005) Final report. An epidemiologic study of mortality and radiation-related risk of cancer among workers at the Idaho National Engineering and Environmental Laboratory, a U.S. Department of Energy facility. NIOSH, Washington |
| 5 | 0.93 | 0.49–1.58 | Ahrenholz S, Cardarelli J, Dill P et al (2001) Final report. Mortality patterns among workers at the Portsmouth gaseous diffusion plant Piketon, Ohio. NIOSH, Washington |
| 5.8 | 0.91 | – | Wilkinson GS, Trieff N, Graham R et al (2000) Final report. Mortality among female nuclear weapons workers. NIOSH, Washington |
| 13.5 | 1.28 | 0.69–2.0 | Boice JD, Cohen SS, Mumma MT et al (2006) Mortality among radiation workers at Rocketdyne (Atomics International), 1948–1999. Radiat Res 166:98–115 |
| 3 | 1.18 | | |
| 7 | 1.11 | | |
| 25 | 1.39 | | |
| 75 | 2.22 | | |
| 800 | 0.22 | – | Spiers FW, Lucas HF, Rundo J et al (1983) Leukaemia incidence in the U.S. dail workers. Health Phys 44(Suppl 1):65–72 |
| 10 | 0.22 | – | Kneale GW, Stewart AM (1993) Comments on Updated analyses of combined mortality data for workers at the Hanford site, Oak Ridge National Laboratory and Rocky Flats Weapons Plant. Radiat Res 136:408–421 |

**Table A11.3** (continued)

| Dose, mSv | RR (SMR) | 90–95% CI | References |
|---|---|---|---|
| 60 | 0.85 | – | Frigerio NA, Stowe RS (1976) Carcinogenic and genetic hazard from background radiation. In: Proceedings of the Biological Effects of Low-Level Radiation Pertinent to Protection of Man and His Environment, Chicago, IL, 1975, IAEA-SM-202/805, vol. 2, International Atomic Energy Agency, Vienna, Austria, pp 385–393 |
| 50 | 0.65 | – | Cohen BL (1995) Test of the linear no-threshold theory of radiation carcinogenesis for inhaled radon decay products. Health Phys 68:157–174 |
| 17.9 | 0.76 | 0.33–1.50 | Rogel A, Carre N, Amoros E et al (2005) Mortality of workers exposed to ionizing radiation at the French National Electricity company. Am J Indust Med 47:72–82 |
| 10,000 | 2.50 | – | Berrington A, Darby SC, Weiss HA et al (2001) 100 years of observation on British radiologists: mortality from cancer and other causes 1897–1997 Br J Radiol 74:507–519 |
| 5,000 | 2.70 | – | |
| 2,000 | 1.75 | – | |
| 800 | 1.16 | – | |
| 7 | 1.09 | 0.58–1.87 | Silver SR, Anderson-Mahoney P, Burphy J et al (2005) Mortality update for the Pantex weapons facility: final report health-related energy research branch. National Institute for Occupational Safety and Health, HHS (NIOSH) Publication No. 2005-124 |
| 5 | 2.03 | 0.78–5.60 | Yiin JH, Schubauer-Berigan MK, Silver SR et al (2005) Risk of lung cancer and leukemia from exposure to ionizing radiation and potential confounders among workers at the Portsmouth Naval Shipyard. Radiat Res 163:603–613 |
| 30 | 3.16 | | |
| 100 | 3.39 | 1.21–8.71 | |
| | | 1.16–9.91 | |
| 21 | 1.01 | 0.73–1.35 | Wiggs LD, Cox-DeVore CA, Voelz GL (1991) Mortality among a cohort of workers monitored for 210Po exposure: 1944–1972. Health Phys 61:71–76 |
| 23.3 | 0.84 | – | Gilbert ES, Omohundro E, Buchanan JA et al (1993) Mortality of workers at the Hanford site: 1945–1986. Health Phys 64:577–590 |

*(continued)*

**Table A11.3** (continued)

| Dose, mSv | RR (SMR) | 90–95% CI | References |
|---|---|---|---|
| 500 | 1.36 | – | Shilnikova NS, Preston DL, Ron E et al (2003) Cancer mortality risk among workers at the Mayak Nuclear Complex. Radiat Res 159:787–798 |
| 1,000 | 1.63 | – | |
| 3,000 | 1.71 | – | |
| 5,000 | 4.09 | – | |
| 7,000 | 10.0 | – | |
| 35 | 0.62 | 0.29–1.23 | Carpenter LM, Higgins CD, Douglas AJ et al (1998) Cancer mortality in relation to monitoring for radionuclide exposure in three UK nuclear industry workforces. Br J Cancer 78:1224–1232 |
| 360 | 1.25 (SIR) | 0.14–4.50 | Ron E, Auvinen A, Alfandary E et al (1999) Cancer risk following radiotherapy for infertility or menstrual disorders. Int J Cancer 82:795–798 |
| 280 | 1.17 | 0.68–2.01 | Franklyn JA, Maisonneuve P, Sheppard M et al (1999) Cancer incidence and mortality after radioiodine treatment for hyperthyroidism: a opoulation-based cohort study. Lancet 353:2111–2115 |
| 60 | 0.94 | 0.7–1.2 | Mohan AK, Hauptmann M, Freedman DM et al (2003) Cancer and other causes of mortality among radiologic technologists in the United States. Int J Cancer 103:259–267 |
| 18.4 | 0.70 | – | Andersson M, Engholm G, Ennow K et al (1991) Cancer risk among staff at two radiotherapy departments in Denmark. Br J Radiol 64:455–460 |
| 3,000 | 2.01 | – | Matanoski GM, Sartwell P, Elliott E et al (1984) Cancer risks in radiologists and radiation workers. In: Boice JD, Fraumeni JF (eds) Radiation carcinogenesis: epidemiology and biological significance. Raven, New York, pp 83–96 |
| 600 | 1.00 | – | |
| 5,000 | 6.15 | – | Smith PG, Doll R (1981) Mortality from cancer and all causes among British radiologists. Br J Radiol 56:187–194 |
| 2,000 | 1.54 | – | |
| 10,000 | 1.55 | – | Yoshinaga S, Aoyama T, Yoshimoto Y et al (1999) Cancer mortality among radiological technologists in Japan: updated analysis of follow-up data from 1969 to 1993. J Epidemiol 9:61–72 |
| 500 | 0.95 | – | |

**Table A11.3** (continued)

| Dose, mSv | RR (SMR) | 90–95% CI | References |
|---|---|---|---|
| 6.3 | 0.82 | 0.59–1.17 | Ashmore JP, Krewski D, Zielinski JM et al (1998) First analysis of mortality and occupational radiation exposure based on the National Dose Registry of Canada. Am J Epidemiol 148:564–574 |
| 35 | 1.09 | 0.87–1.32 | Sigurdson AJ, Doody MM, Rao RS et al (2003) Cancer incidence in the U.S. radiologic technologists health study, 1983–1998. Cancer 97:3080–3089 |
| 180 | 1.25 | – | Miller RW, Jablon S (1970) A search for late radiation effects among men who served as X-ray technologists in the U.S. Army during World War II. Radiology 96:269–274 |
| 10 | 0.76 | 0.32–1.26 | Telle-Lamberton M, Bergot D, Gagneau M et al (2004) Cancer mortality among French Atomic Energy Commission workers. Am J Ind Med 45:34–44 |
| 30 | 0.35 | – | Wiggs LD, Johnson ER, Cox-DeVore CA et al (1994) Mortality through 1990 among white male workers at the Los Alamos National Laboratory: considering exposures to plutonium and external ionizing radiation. Health Phys 67:577–588 |
| 75 | 0.67 | – | |
| 140 | 1.26 | – | |
| 48 | 1.11 | 0.57–1.89 | Dupree-Ellis E, Watkins J, Ingle JN et al (2000) External radiation exposure and mortality in a cohort of uranium processing workers. Am J Epidemiol 152:91–95 |
| 4,000 | 3.2 (SIR) | 1.5–6.1 | Shore RE, Moseson M, Harley N et al (2003) Tumors and other diseases following childhood X-ray treatment for ringworm of the scalp (*Tinea capitis*). Health Phys 85:404–408 |
| 52 | 0.60 | – | Gribbin MA, Weeks JL, Howe GR (1993) Cancer mortality (1956–1985) among male employees of Atomic Energy of Canada Limited with respect to occupational exposure to external low-linear-energy transfer ionizing radiation. Radiat Res 133:375–380 |
| 13.2 | 0.95 | – | Frome EL, Cragle DL, Watkins JP et al (1997) A mortality study of employees of the nuclear industry in Oak Ridge, Tennessee. Radiat Res 148:64–80 |
| 11.4 | 1.01 | – | Artalejo FR, Lara SC, de Andres Manzano B et al (1997). Occupational exposure to ionizing radiation and mortality among workers of the former Spanish Nuclear Energy Board. Occup Environ Med 54:202–208 |

*(continued)*

**Table A11.3** (continued)

| Dose, mSv | RR (SMR) | 90–95% CI | References |
|---|---|---|---|
| 15 | 0.74 | – | Beral V, Fraser P, Carpenter L et al (1988) Mortality of employees of the Atomic Weapons establishment, 1951–1982. BMJ 297:757–770 |
| 30.5 | 0.91 | – | Muirhead CR, Goodill AA, Haylock RGE et al (1999) Occupational radiation exposure and mortality: second analysis of the National Registry of Radiation Workers. J Radiol Prot 19:3–26 |
| 87.4 | 0.71 | – | Omar RZ, Barber JA, Smith PG (1999) Cancer mortality and morbidity among plutonium workers at the Sellafield plant of British Nuclear Fuels. Br J Cancer 79:1288–1301 |
| 30.7 | 1.19 | – | Fraser P, Carpenter L, Maconochie N et al (1993) Cancer mortality and morbidity in employees of the United Kingdom Atomic Energy Authority, 1946–1986. Br J Cancer 67:615–624 |
| ?? | 0.90 | – | Reynolds P, Austin DF (1985) Cancer incidence among employees of the Larence Livermore National Laboratory. West J Med 142:214–218 |
| 15.2 | 1.63 | – | Wing S, Shy CM, Wood JL et al (1991) Mortality among workers at Oak Ridge National Laboratory. Evidence of radiation effects in follow-up through 1984. JAMA 265:1397–1402 |
| 13.2 | 0.60 | – | Loomis DP, Wolf SH (1996) Mortality of workers at a nuclear materials production plant at Oak Ridge, Tennessee, 1947–1990. Am J Ind Med 29:131–141 |
| 10 | 0.62 | 0.07–2.24 | Lim YK, Yoo KY (2006) A cohort study on cancer risk by low-dose radiation exposure among radiation workers of nuclear power plants in Korea. J Korea Assoc Radiat Prot 31:53–63 |
| 5 | 0.99 | 0.60–1.54 | Iwasaki T, Murata M, Ohshima S et al (2003) Second analysis of mortality of nuclear industry workers in Japan, 1986–1997. Radiat Res 159:228–238 |
| 15 | 0.72 | | |
| 35 | 1.57 | 0.09–2.60 | |
| 75 | 0.58 | | |
| 135 | 0.93 | 0.51–3.66 | |
| | | 0.02–3.22 | |
| | | 0.02–5.20 | |

**Table A11.3** (continued)

| Dose, mSv | RR (SMR) | 90–95% CI | References |
|---|---|---|---|
| 47.8 | 2.1 | 0.8–4.3 | Hwang SL, Guo HR, Hsieh WA et al (2006) Cancer risks in a population with prolonged low-dose-rate γ-radiation exposure in radiocontaminated buildings, 1983–2022. Int J Radiat Biol 82:849–858 |
| 29 | 1.16 | 0.46–2.87 | Zeeb H, Blettner M, Langner I et al (2003) Mortality from cancer and other causes among airline cabin attendants in Europe: a collaborative cohort study in eight countries. Am J Epidemiol 158:35–46 |
| 30 | 1.20 | – | Lynge E (2001) Commentary: cancer in the air. Int J Epidemiol 30:830–832 |
| | | – | |
| 10 | 1.08 | 1.01–1.16 | Kubale TL, Daniels RD, Yiin JH et al (2005) A nested case-control study of leukemia mortality and ionizing radiation at the Portsmouth Naval Shipyard. Radiat Res 164:810–819 |
| 1,720 | 1.80 | 1.1–2.8 | Curtis RE, Boice JD, Stovall M et al (1994) Relationship of leukemia risk to radiation dose following cancer of the uterine corpus. J Natl Cancer Inst 86:1315–1324 |
| 9,880 | 2.29 | 1.4–3.7 | |
| 5 | 1.08 | 0.65–1.84 | Schubauer-Berigan MK, Daniels RD, Fleming DA et al (2007) Risk of chronic myeloid and acute leukemia mortality after exposure to ionizing radiation among workers at four U.S. nuclear weapons facilities and a nuclear naval shipyard. Radiat Res 167:222–232 |
| 30 | 1.68 | | |
| 75 | 2.09 | 0.99–2.91 | |
| 150 | 2.54 | | |
| | | 1.00–4.34 | |
| | | 1.22–5.26 | |

## References


1. Lichtman MA, Klemperer MR (1978) Clinical oncology. University of Rochester, Rochester, NY and American Cancer Society, Atlanta, p 245
2. Lewis E. (1957) Leukemia and ionizing radiation. Science 43:965
3. Rosner F, Grunwald HW, Zarrabi MH (1979) Acute leukemia as a complication of cytotoxic chemotherapy. Int J Radiat Oncol Biol Phys 5:1705–1707

4. Boivin JF, Hutchison GB (1981) Leukemia and other cancers after radiotherapyand chemotherapy for Hodgkin's disease. JNCI 67:751–760
5. Committee to Assess Health Risks from Exposure to Low Levels of Ionizing Radiation (2007) Health risks from exposure to low levels of ionizing radiation: BEIR VII – phase 2. National Academies Press, Washington, DC
6. Sanders CL (2006) Hormesis as a confounding factor in epidemiological studies of radiation carcinogenesis. Korean Assoc Radiat Prot 31:69–89
7. Robinette C, Jablob S, Preston TL (1985) Studies of participants in nuclear tests. Final Report DOE/EV/0157. National Research Council, Washington, DC
8. Sever LE, Gilbert ES (1997) Epidemiologic evaluation of childhood leukemia and paternal exposure to ionizing radiation. Final Report, Centers for Disease Control and Prevention U50/CCU012545–01. Battelle Memorial Institute, Seattle, WA
9. Doll R, Wakeford R (1997) Risk of childhood cancer from fetal irradiation. Br J Radiol 70: 130–139
10. Jaumburg E, Bellecco R, Cnattingius S et al (2002) Intrauterine exposure to diagnostic X rays and risk of childhood leukemia subtypes. Radiat Res 156:718–723
11. Shu XO, Potter JD, Linet MS et al (2002) Diagnostic X rays and ultrasound exposure and risk of childhood acute lymphoblastic leukemia by immunophenotype. Cancer Epidemiol Biomarkers Prev 11:177–185
12. Rodvall Y, Hrubec Z, Pershagen G et al (1992) Childhood cancer among Swedish twins. Cancer Causes Control 3:527–532
13. Delongchamp RR, Mabushi K, Yasuhiko Y et al (1997) Cancer mortality among atomic bomb survivors exposed *in utero* or as young children. Radiat Res 147:385–395
14. IARC (2000) Monograph on the evaluation of carcinogenic risks to humans, vol 75, Ionizing radiation part I: X and gamma radiation and neutrons. Lyon, France
15. Folley JH, Borges W, Yamasaki TY (1952) Incidence of leukemia in survivors of the atom bomb in Hiroshima and Nagasaki, Japan. Am J Med 13:311–321
16. Preston DL, Kusumi S, Tomonaga M et al (1994) Cancer incidence in atomic bomb survivors. Part III. Leukemia, lymphoma and multiple myeloma, 1950–1987. Radiat Res 137:S68–S97
17. UNSCEAR (1958) Report of the United Nations Scientific Committee on the Effects of Atomic Radiation. United Nations, New York, pp 1–228
18. Shimizu Y, Kato H, Schull WJ (1990) Studies on the mortality of A-bomb survivors. 9, Mortality, 1950–1985: Part 2, Cancer mortality based on the recently revised doses (DS86). Radiat Res 121:120–141
19. Mine M, Okumura Y, Ichimara M et al (1990). Apparently beneficial effect of low to intermediate doses of A-bomb radiation on human lifespan. Int J Radiat Biol 58:1035–1043
20. Land CE (1980) Estimating cancer risks from low doses of ionizing radiation. Science 209: 1197
21. Viel JF (1993) Radon exposure and leukemia in adulthood. Int J Epidemiol 22:627–630
22. Richardson C, Monfort M, Green M et al (1995) Spatial variation of natural radiation and childhood leukemia incidence in Great Britain. Stat Med 14:2487–2501
23. Craig L, Seidman H (1961) Leukemia and lymphoma mortality in relation to cosmic radiation. Blood 17:319
24. Evard A-S, Hemon D, Billon S et al (2006) Childhood leukemia incidence and exposure to indoor radon, terrestrial and cosmic gamma radiation. Health Phys 90:569–579
25. Evrard AS, Hemon D, Billon S et al (2005) Ecological association between indoor radon concentration and childhood leukemia incidence in France, 1990–1998. Eur J Cancer Prev 14: 147–157
26. Laurier D, Valenty M, Tirmarche M (2001) Radon exposure and the risk of leukemia: a review of epidemiological studies. Health Phys 81:272–288
27. Yoshinaga S, Tokonami S, Akiba S (2005) Residential radon and childhood leukemia: a meta-analysis of published studies. Int Congress Ser 1276:430–431

28. Canu IG, Ellis ED, Tirmarche M (2008) Cancer risk in uranium workers occupationally exposed to uranium-emphasis on internal exposure. Health Phys 94:1–17
29. Darby SC, Whitley E, Howe GR et al (1995) Radon and cancers other than lung cancer in underground miners: a collaborative analysis of 11 studies. J Natl Cancer Inst 87:378–384
30. Xiao WU, Jiang R, Chang X et al (2006) Epidemiologist investigate on mortality of uranium miner in Jiangxi Province. In: Proceedings of the Second Asian and Oceanic Congress for Radiation Protection, 9–13 October 2006, Beijing, China, pp 1314–1318
31. Veiga LHS, Amaral ECS, Colin D, Koifman S (2006) A retrospective mortality study of workers exposed to radon in a Brazilian underground coal mine. Radiat Environ Biophys 45: 125–134
32. Ostroumova E, Gagniere B, Lauier D et al (2006) Risk analysis of leukaemia incidence among people living along the Techa River: a nested case-control study. J Radiol Prot 26:17–32
33. Krestinina L Yu, Preston DL, Ostroumova EV et al (2005) Protracted radiation exposure and cancer mortality in the Techa River cohort. Radiat Res 164:602–611
34. Soloviev VYu, Semenov VG, Koshurnikova NA et al (2006) 'Early' leukaemia effect in prolonged exposure with high doses. Intern J Low Radiat 2:275–284
35. Vaughan J (1973) Handbook of experimental pathology, Vol 36: Uranium, plutonium, transplutonic elements. Springer, Berlin, p 349
36. Sanders CL (1996) Prevention and therapy of cancer and other common diseases: alternative and traditional approaches. Infomedix, Richland, WA, 3000pp
37. Baverstock KF, Papworth D (1989) The UK radium luminizer survey. Br J Radiol 21:71–76
38. Spiers FW, Lucas HF, Rundo J et al (1983) Leukemia incidence in the U.S. dial workers. Health Phys 44(Suppl 1):65–72
39. Hatch M, Ron E, Bouville A et al (2005) The Chernobyl disaster: cancer following the accident at the Chernobyl nuclear power plant. Epidemiol Rev 27:56–66
40. Chernobyl Forum 2003–2005. (2006) Chernobyl legacy: health, environmental and socioeconomic impacts. IAEA, 57 pages
41. Bol'shov LA, Gabaraev BA, It'in LA et al (2000) Comparison of accident risks in different energy systems: comments from Russian specialists. IAEA Bull 42(4)
42. Bennett B, Repacholi M, Carr Z (eds) (2006) Health effects of the chernobyl accident and special health care programmes. World Health Organization, Geneva, pp 57–60
43. Ivanov VK (2007) Late cancer and noncancer risks among Chernobyl emergency workers of Russia. Health Phys 93:470–479
44. Howe GR, Zablotska LB, Fix JJ et al (2004) Analysis of the mortality experience amongst U.S. nuclear power industry workers after chronic low-dose exposure to ionizing radiation. Radiat Res 162:517–526
45. Zablotska LB, Ashmore JP, Howe GR (2004) Analysis of mortality among Canadian nuclear power industry workers after chronic low-dose exposure to ionizing radiation. Radiat Res 161:633–641
46. Beral V, P Fraser, Both M et al (1987) Epidemiological studies of workers in the nuclear industry. In: Jones RR, Southwood R (eds) Radiation & health. Wiley, New York, pp 97–106
47. Cardis E, Gilbert ES, Carpenter L et al (1995) Effects of low doses and low dose rates of external ionizing radiation: cancer mortality among nuclear industry workers in three countries. Radiat Res 142:117–132
48. Vrijheid M, Cardis E, Ashmore P et al (2008) Ionizing radiation and risk of chronic lymphocytic leukemia in the 15-country study of nuclear industry workers. Radiat Res 170:661–665

# Liver, CNS, and Thyroid Cancers 12

The very existence of radiation hormesis phenomenon proves the existence of radiation thresholds and falsifies LNT. This is why radiation hormesis is the best remedy for mass psychological affliction called radiophobia, and, by the same token, this is why it is ignored by the influential part of the radiation protection establishment, against a vast factual evidence and the benefit of society [1].

## 12.1 Liver Cancer

Liver cancer is one of the most frequent cancers in the world with a large geographical variation in frequency. Liver cancer is one of the most prevalent cancers in South Korea and Japan. It is the third most prevalent cancer among Korean men (15%) and seventh most prevalent cancer in Korean women (6%). Prevalence of liver cancer in the United States is much less for both genders.

Confusion may result in differentiating primary liver cancer from metastatic liver cancer. Risk factors for primary liver cancer include hepatitis B and C infections, alcoholism, aflatoxin $B_1$, other mycotoxins, tamoxifen, liver fluke, pyrrolizidine alkaloids from certain plants, vinyl chloride, tobacco-specific nitrosamines, heterocyclic aromatic amines, Thorotrast, hemochromatosis, and inherited genetic disease (alpha-1-antitrypsin deficiency, tyrosinemia, Wilson's disease, and glycogen storage disease) [2, 3]. The majority of primary liver cancers in Japan are associated chronic viral hepatitis infections [4]. Spontaneous rates of primary liver cancer are as much as three times higher in males than in females. The majority of primary liver cancers are hepatocellular carcinomas, followed by cholangiocarcinomas.

In Thorotrast patients, liver cancers are dominated by cholangiocarcinomas and hemangiosarcomas [5]. Thorotrast is colloidal $^{232}ThO_2$ that was used as an intravenously injected X-ray contrast medium from 1928 to 1955. Thorotrast aggregates primarily in the spleen and liver. Some daughter products, particularly $^{224}Ra$, redeposit in the bone. A typical 25-mL injection of Thorotrast in an adult gives an average dose rate to the liver of 250 mGy year$^{-1}$, while the average endosteal dose rate from daughter products and Thorotrast is 180 mGy year$^{-1}$. Liver cancer is the most common tumor (6% of Thorotrast patients) followed by leukemia [6, 7].

C. L. L. Sanders, *Radiation Hormesis and the Linear-No-Threshold Assumption*,
DOI: 10.1007/978-3-642-03720-7_12, © Springer Verlag Berlin Heidelberg 2010

**Table 12.1** Relative risk of liver tumors in Mayak plutonium workers. In addition to α-dose in the liver, workers also received an average whole-body dose of ~1 Gy from external $^{60}$Co γ-rays [9]

| α-Dose range (Gy) | RR | 95% CI |
|---|---|---|
| 0–0.07 | 1.0 | – |
| >0.07–0.54 | 0.40 | 0.14–1.12 |
| >0.54–16.9 | 2.44 | 1.09–5.44 |

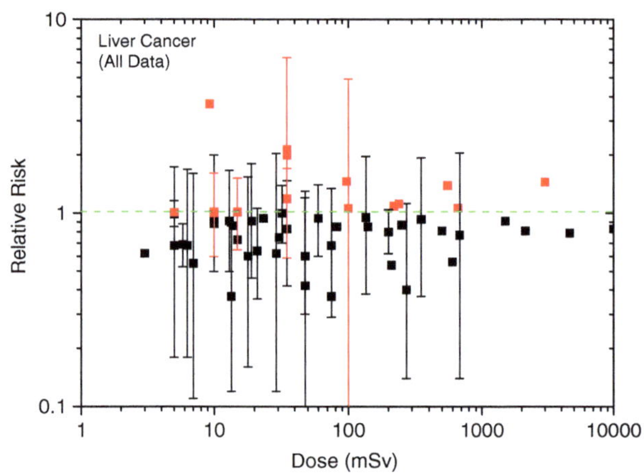


**Fig. 12.1** Liver cancer risk and radiation dose in human populations exposed to ionizing radiation

A threshold of ~2 Gy to the liver was associated with a significantly increase incidence of primary liver tumors in both Thorotrast patients [8] and in plutonium workers [9]. The risk of liver cancer in plutonium workers at liver doses of about 0.1–0.5 Gy was 60% less than expected (Table 12.1) [9]. Liver cancer risk was evaluated in 48 published studies of irradiated populations. About 70% of the liver tumor-dose data points exhibited RR < 1.0, even up to liver doses of 10 Sv. No dose–response relationship was noted (Appendix Table A12.3; Fig. 12.1).

## 12.2 Central Nervous System Cancer

Primary brain tumor incidence in the U.S. has increased about 1% per year among all age groups and 3% per year among persons >65 years of age [10]. Tumors of the brain are classified as to their origins: neuroectodermal cells (gliomas derived from astrocytes-astrocytoma, oligodendroglia-oligodendroglioma, and ependyma-ependymoma, mesenchymal cells) (meningioma, schwannoma), cells ectopically displaced during embryogenesis (craniopharyngioma), gliosarcoma, and metastatic tumors. Gliosarcoma often arises following radiotherapy of glioblastoma multiforme as a variant of this tumor [11].

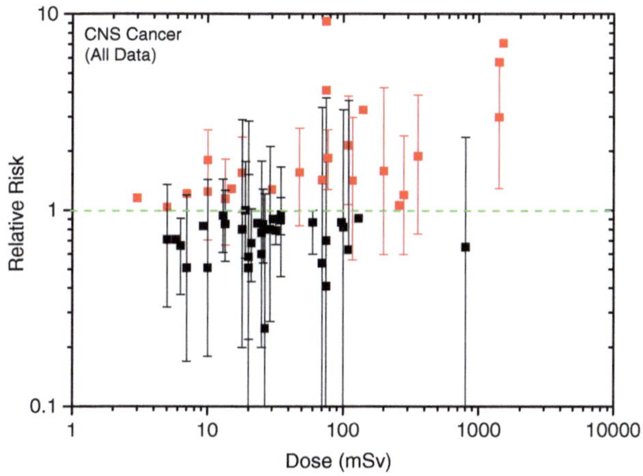


**Fig. 12.2** CNS tumor risk and radiation dose in human populations exposed to ionizing radiation

An increased mortality from brain tumors (astrocytoma and meningioma) has been associated in humans exposed to toxic chemicals (organic solvents, halomethanes, and chlorinated aliphatic hydrocarbons), scars from head injuries, and cranial irradiation from a variety of sources. Genetic syndromes, such as hereditary neurofibromatosis and Li-Fraumeni syndrome familial cancer, are consistently linked with a high risk of primary brain tumors [3].

High-dose irradiation of the brain during childhood is associated with increased brain tumor frequency [12]. An increased mortality from brain tumors was not observed in the Japanese A-bomb survivors [13]. Nonsignificant increases in brain tumors has been found in Thorotrast patients [14] and in plutonium workers [15, 16]. A nonsignificant increase in brain tumors was found in studies of inhaled plutonium in rats [17]. Glioblastomas of the brain developed in 9 of 11 Rhesus monkeys within ten years after receiving ten fractions of 3.5 Gy to the whole brain [18]. Increased malignant and benign brain tumors are seen after exposure of the whole brain to ~1.5 Gy during scalp therapy for tinea capitis [12].

Central nervous system (CNS) tumor risk was evaluated in 45 published studies of irradiated populations. RR < 1.0 predominated at cumulative doses <70 mSv. A possible increase in CNS tumors was seen at doses >100 mSv (Appendix Table A12.4; Fig. 12.2).

## 12.3 Thyroid Cancer

A U.S. National Council on Radiation Protection report on thyroid cancer said, "available human data on low dose I-131 exposures have not shown I-131 to be carcinogenic in the human thyroid" [22]. No increase in thyroid cancer was found in the United States after hundreds of A-bomb tests in Nevada during the 1950s and 1960s [23, 24]. However, thyroid cancer incidence in children is correlated with radiation dose [25], with an apparent threshold of 200 mSv [26].

**Table 12.2** Influence of age at exposure on the relative risk of thyroid cancer from a dose of 1 Gy [19, 20]. [thyroid gland weight is 1 g (birth), 2 g (6 months), 4 g (4 years), 20g (10+ years)] [21]

| External radiation of patients | | Japanese A-bomb survivors | |
|---|---|---|---|
| Age | RR | Age | RR |
| 0–4 years | 40 | 0–9 years | 9.5 |
| 5–9 years | 20 | 10–19 years | 3 |
| 10–14 years | 10 | 20–39 years | 0.3 |
| 20–30 years | 1 | >40 years | 0.2 |
| >30 years | 0 | | |

High doses of thyroid irradiation increase the risk for papillary and follicular thyroid carcinomas from 15- to 53-fold, with the risk being the highest following radiation at a young age. Clear evidence of radiation hormesis is seen in those irradiation at an age >20 years (Table 12.2) [19, 20, 27]. A four- to fivefold increase in thyroid cancer risk was found in children receiving radiotherapy for tinea capitis [28, 29].

Thyroid screening programs are effective in identifying occult thyroid cancers [30–32]. The incidence of dormant, occult tumors is common and variable among countries. For example, the occult thyroid tumor incidence is 9% in Poland and 36% in Finland [33].

The increased thyroid tumor incidence following irradiation from the Chernobyl accident can be mostly attributed to improved screening and reporting. A screening program in the U.S. several years before 1986 resulted in a sevenfold increased in diagnosed thyroid cancers and a 17-fold increase in thyroid nodules, similar to what was observed in Belarus following the Chernobyl accident [34]. Many published studies of thyroid cancer following the Chernobyl accident have been ecological studies [35]. From 17–25% of contaminated residents in Ukraine and Belarus, mostly young children, accumulated thyroid doses >1 Gy [36]. Several thousand cases of thyroid tumors, mainly in those who were exposed as children and adolescents, have been found in the populations probably due to markedly increased thyroid screening programs; only nine persons have died of thyroid cancer [36, 37].

Iodine deficiency increases both thyroid mass and fractional uptake of $^{131}$I by the thyroid. Excess relative risk estimates of thyroid cancer would be biased upward if there is a correlation between thyroid dose and frequency of screening. Much is unknown in the Chernobyl exposed populations about the thyroid cancer dose response due to uncertainties in dose estimation and screening biases [35].

Iodine-131 used in the diagnosis of thyroid disease resulted in a decrease in thyroid cancer from expected incidence [38–40]. About 35,000 Swedish patients who received a thyroid dose of 0.5 Gy from $^{131}$I for diagnostic purposes experienced a 38% reduction in expected thyroid cancers [41]. Negative excess relative risks for nonthyroid cancers have been observed in patients treated with $^{131}$I for hyperthyroidism or thyroid cancer with ERR values ranging from −0.07 to −0.27 [42, 43]. The Hanford Thyroid Disease Study showed a negative slope for thyroid cancer risk vs. dose to the thyroid [44].

Thyroid cancer risk was evaluated in 31 published studies of irradiated populations. The thyroid cancer-dose data points for all exposed populations appeared somewhat

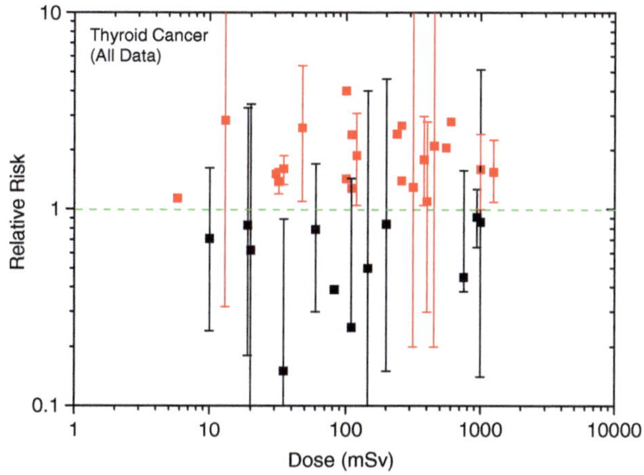


**Fig. 12.3** Thyroid cancer risk and radiation dose in human populations exposed to ionizing radiation

uniformly distributed above and below the RR = 1.0 line up to cumulative doses of 1,000 mSv. No dose–response relationship was noted (Appendix Table A12.5; Fig. 12.3).

## Appendix

**Table A12.3** Risk of liver cancer in epidemiological studies of populations exposed to ionizing radiation

| Dose (mSv) | RR (SMR) | 90–95% CI | Reference |
|---|---|---|---|
| 200 | 0.80 | 0.62–1.04 | Tao Z, Zha Y, Akiba S et al (2000) Cancer mortality in the high background radiation areas of Yangjiang, China during the period between 1979 and 1995. J Radiat Res 41(Suppl):31–41 |
| 5 | 0.68 | 0.18–1.73 | Ahrenholz S, Cardarelli J, Dill P et al (2001) Final report. Mortality patterns among workers at the Portsmouth gaseous diffusion plant Piketon, Ohio. NIOSH |
| 13.5<br>3 | 0.37<br>0.62 | 0.12–0.86 | Boice JD, Cohen SS, Mumma MT et al (2006) Mortality among radiation workers at Rocketdyne (Atomics International), 1948–1999. Radiat Res 166:98–115 |
| 13 | 0.91 | 0.50–1.66 | Schubauer-Berigan MK, Macievic GV, Utterback DF et al (2005) Final report. An epidemiologic study of mortality and radiation-related risk of cancer among workers at the Idaho National Engineering and Environmental Laboratory, a U.S. Department of Energy facility. NIOSH |

(*continued*)

**Table A12.3** (continued)

| Dose (mSv) | RR (SMR) | 90–95% CI | Reference |
|---|---|---|---|
| 7 | 0.55 | 0.11–1.60 | Silver SR, Anderson-Mahoney P, Burphy J et al (2005) Mortality update for the Pantex weapons facility: Final report health-related energy research branch. National Institute for Occupational Safety and Health, HHS (NIOSH) Publication No. 2005–124 |
| 17.9 | 0.60 | 0.16–1.54 | Rogel A, Carre N, Amoros E et al (2005) Mortality of workers exposed to ionizing radiation at the French National Electricity company. Am J Indust Med 47:72–82 |
| 678 (lung) | 0.77 | 0.56–1.04 | Wu X, Jiang R (2006) Epidemiology investigate on mortality of uranium miner in Jiangxi Province. Abstracts of The Second Asian and Oceanic Congress for Radiation Protection, Beijing, China, p 354; Wu X (2006) Personal Communication |
| 140 | 0.85 | – | Ye W, Sobue T, Lee VS et al (1998) Mortality and cancer incidence in Misasa, Japan, a spa area with elevated radon levels. Jpn J Cancer Res 89:789–796 |
| 10 | 0.88 | – | McGeoghegan D, Binks K (2000) The mortality and cancer morbidity experience of workers at the Capenhurst uranium enrichment facility 1946–95. J Radiol Prot 20:381–401 |
| 48 | 0.6 (SIR) | 0.3–1.2 | Hwang SL, Guo HR, Hsieh WA et al (2006) Cancer risks in a population with prolonged low dose-rate γ-radiation exposure in radiocontaminated buildings, 1983–2002 Int J Radiat Biol 82:849–858 |
| 19 | 0.91 | 0.46–1.80 | Atkinson WD, Law DV, Bromley KJ et al (2004) Mortality of employees of the United Kingdom Atomic Energy Authority, 1946–97. Occup Environ Med 61:577–585 |
| 5 | 1.01 | – | |
| 15 | | – | |
| 35 | 0.73 | – | |
| 75 | 2.13 | – | |
| | 0.37 | | |
| 551 | 1.39 | | Wang, JX, Zhang LA, Li BX et al (2002) Cancer incidence and risk estimation among medical X-ray workers in China, 1950–1995. Health Phys 82:455–466 |
| 82 | 0.85 | | |
| 97 | 1.46 | – | Sponsler R, Cameron JR (2005) Nuclear shipyard worker study (1980–1988): a large cohort exposed to low-dose-rate gamma radiation. Int J Low Radiation 1:463–478 |
| 100 | 1.06 | 0.05–4.94 | Blettner M, Zeeb H, Langner I et al (2002) Mortality from cancer and other causes among airline cabin attendants in Germany, 1960–1997. Am J Epidemiol 156:556–565 |

**Table A12.3** (continued)

| Dose (mSv) | RR (SMR) | 90–95% CI | Reference |
|---|---|---|---|
| 32 | 1.00 (SIR) | 0.70–1.39 | Sont WN, Zielinski JM, Ashmore JP et al (2001) First analysis of cancer incidence and occupational radiation exposure based on the National Dose Registry of Canada. Am J Epidemiol 153:309–318 |
| 21 | 0.64 | 0.36–1.06 | Wiggs LD, Cox-DeVore CA, Voelz GL (1991) Mortality among a cohort of workers monitored for $^{210}$Po exposure: 1944–1972. Health Phys 61:71–76 |
| 23.3 | 0.94 | – | Gilbert ES, Omohundro E, Buchanan JA et al (1993) Mortality of workers at the Hanford site: 1945–1986. Health Phys 64:577–590 |
| 5.8 | 0.69 | 0.53–0.88 | Wilkinson GS, Trieff N, Graham R et al (2000) Final report. Mortality among female nuclear weapons workers. NIOSH |
| 35 | 2.00 | 0.59–6.38 | Carpenter LM, Higgins CD, Douglas AJ et al (1998) Cancer mortality in relation to monitoring for radionuclide exposure in three UK nuclear industry workforces. Br J Cancer 78:1224–1232 |
| 9.3 | 3.68 | – | McGeoghegan D, Binks K (2000) The mortality and cancer morbidity experience of workers at the Springfields uranium production facility, 1946–95. J Radiol Prot 20:111–137 |
| α–mGy<br>10<br>273<br>540–16,900 | 1.0<br>0.40<br>2.44 | –<br>0.14–1.12<br>1.09–5.44 | Tokarskaya ZB, Zhuntova GV, Scott BR et al (2006) Influence of alpha and gamma radiations and non-radiation risk factors on the incidence of malignant liver tumors among Mayak PA workers. Health Phys 91:296–310 |
| 350 | 0.93 | 0.37–1.93 | Tirmarche M, Raphalen A, Allin F et al (1993) Mortality of a cohort of French uranium miners exposed to relatively low radon concentrations. Br J Cancer 67:1090–1097 |
| 60 | 0.94 | 0.6–1.4 | Mohan AK, Hauptmann M, Freedman DM et al (2003) Cancer and other causes of mortality among radiologic technologists in the United States. Int J Cancer 103:259–267 |
| 3,000<br>600 | 1.45<br>0.56 | –<br>– | Matanoski GM, Sartwell P, Elliott E et al (1984) Cancer risks in radiologists and radiation workers. In: Boice JD, Fraumeni JF (eds). Radiation Carcinogenesis: epidemiology and biological significance. Raven, New York, 83–96 |
| 10,000<br>500 | 0.83<br>0.81 | –<br>– | Yoshinaga S, Aoyama T, Yoshimoto Y et al (1999) Cancer mortality among radiological technologists in Japan: updated analysis of follow-up data from 1969 to 1993. J epidemiol 9:61–72 |

*(continued)*

**Table A12.3** (continued)

| Dose (mSv) | RR (SMR) | 90–95% CI | Reference |
|---|---|---|---|
| 35 | 0.83 | 0.42–1.47 | Sigurdson AJ, Doody MM, Rao RS et al (2003) Cancer incidence in the U.S. radiologic technologists health study, 1983–1998. Cancer 97:3080–3089 |
| 10 | 0.95 | 0.50–1.99 | Telle-Lamberton M, Bergot D, Gagneau M et al (2004) Cancer mortality among French Atomic Energy Commission workers. Am J Indust Med 45:34–44 |
| 30 | 0.0 | – | Wiggs LD, Johnson ER, Cox-DeVore CA et al (1994) Mortality through 1990 among white male workers at the Los Alamos National Laboratory: considering exposures to plutonium and external ionizing radiation. Health Phys 67:577–588 |
| 75 | 0.0 | – | |
| 140 | 0.0 | – | |
| 48 | 0.42 | 0.07–1.30 | Dupree-Ellis E, Watkins J, Ingle JN et al (2000) External radiation exposure and mortality in a cohort of uranium processing workers. Am J Epidemiol 152:91–95 |
| 6.3 | 0.68 | 0.18–1.68 | Ashmore JP, Krewski D, Ziellnski JM et al (1998) First analysis of mortality and occupational radiation exposure based on the National Dose Registry of Canada. Am J Epidemiol 148:564–574 |
| 30.5 | 0.75 | – | Muirhead CR, Goodill AA, Haylock RGE et al (1999) Occupational radiation exposure and mortality: second analysis of the National Registry of Radiation Workers. J Radiol Prot 19:3–26 |
| 10 | 1.02 | 0.60–1.61 | Lim YK, Yoo KY (2006) A cohort study on cancer risk by low-dose radiation exposure among radiation workers of nuclear power plants in Korea. J Korea Assoc Radiat Prot 31:53–63 |
| 5 | 0.99 | 0.85–1.16 | Iwasaki T, Murata M, Ohshima S et al (2003) Second analysis of mortality of nuclear industry workers in Japan, 1986–1997. Radiat Res 159:228–238 |
| 15 | 1.02 | 0.65–1.52 | |
| 35 | 1.19 | 0.80–1.71 | |
| 75 | 0.68 | 0.29–1.34 | |
| 135 | 0.95 | 0.38–1.96 | |
| 240 | 1.12 (SIR) | – | Thompson DE, Mabuchi K, Ron E et al (1994) Cancer incidence in atomic bomb survivors. Part II. Solid tumors, 1958–1987. Radiat Res 137:S17–S67 |
| 1,500 | 0.91 (SIR) | – | Boice JD, Day NE, Andersen A et al (1985) Second cancers following radiation treatment for cervical cancer. An international collaboration among cancer registries. J Natl Cancer Inst 74:955–975 |

**Table A12.3** (continued)

| Dose (mSv) | RR (SMR) | 90–95% CI | Reference |
|---|---|---|---|
| 660 | 1.06 (SIR) | – | Mattsson A, Hall P, Ruden BI et al (1997) Incidence of primary malignancies other than breast cancer among women treated with radiation therapy for benign breast disease. Radiat Res 148:152–160 |
| 220 | 1.09 | – | Pierce DA, Shimizu Y, Preston DL et al (1996) Studies of the mortality of A-bomb survivors. Report 12, Part I. Cancer: 1950–1990. Radiat Res 146:1–27 |
| 2,130 | 0.81 | – | Weiss HA, Darby SC, Doll R (1994) Cancer mortality following x-ray treatment for ankylosing spondylitis. Int J Cancer 59:327–338 |
| 4,610 | 0.79 | – | Griem ML, Kleinerman RA, Boice JD et al (1994) Cancer following radiotherapy for peptic ulcer. J Natl Cancer Inst 86:842–849 |
| 210 | 0.54 | – | Inskip PD, Monson RR, Wagoner JK et al (1990) Cancer mortality following radium treatment for uterine bleeding. Radiat Res 123:331–344 |
| 14 | 0.86 | – | Epidemiological Study Group of Nuclear Workers (Japan) (1997) First analysis of mortality of nuclear industry workers in Japan, 1986–1992. J Health Phys 32:173–184 |
| 250 | 0.87 | – | Ron E, Doody MM, Becker DV et al (1998) Cancer mortality following treatment for adult hyperthyroidism. JAMA 280:347–355 |
| 45,000 Q = 10 | 120 | – | Andersson M, Carstensen B, Storm HH (1995) Mortality and cancer incidence after cerebral arteriography with or without Thorotrast. Radiat Res 142:305–320 |
| 49,000 Q = 10 | 126 | – | Van Kaick G, Dalheimer A, Hornik S et al (1999) The German Thorotrast study: recent results and assessment of risks. Radiat Res. 152:S64–S71 |
| 30,000 Q = 10 | 15.8 | – | Dos Santos Silva I, Malveiro F, Portugal R et al (1995) Mortality from primary liver cancers in the Portuguese Thorotrast cohort study. In: van Kaick G et al (eds) Health effects of internally deposited radionuclides: emphasis on radiation and thorium. World Scientific, Singapore, pp 229–233 |
| 47.8 | 0.6 | 0.3–1.2 | Hwang SL, Guo HR, Hsieh WA et al (2006) Cancer risks in a population with prolonged low-dose-rate γ-radiation exposure in radiocontaminated buildings, 1983–2022. Int J Radiat Biol 82:849–858 |

(*continued*)

**Table A12.3** (continued)

| Dose (mSv) | RR (SMR) | 90–95% CI | Reference |
|---|---|---|---|
| 29 | 0.62 | 0.12–2.03 | Zeeb H, Blettner M, Langner I et al (2003) Mortality from cancer and other causes among airline cabin attendants in Europe: a collaborative cohort study in eight countries. Am J Epidemiol 158:35–46 |
| 678 | 0.77 | 0.14–2.04 | Xiao WU, Jiang R, Chang X et al (2006) Epidemiologist investigate on mortality of uranium miner in Jiangxi Province. Proceedings of The Second Asian and Oceanic Congress for Radiation Protection, October 9–13, 2006, Beijing, China, pp 1314–1318 |

**Table A12.4** Risk of CNS cancer in epidemiological studies of populations exposed to ionizing radiation

| Dose (mSv) | RR (SMR) | 90–95% CI | Reference |
|---|---|---|---|
| 200 | 1.59 | 0.60–4.23 | Tao Z, Zha Y, Akiba S et al (2000) Cancer mortality in the high background radiation areas of Yangjiang, China during the period between 1979 and 1995. J Radiat Res 41(Suppl):31–41 |
| 20 | 0.51 | 0.01–2.86 | Pukkala E, Auvinen A, Wahlberg G (1995) Incidence of cancer among Finnish airline cabin attendants, 1967–92. BMJ 311:649–652 |
| 5 | 0.71 | 0.32–1.35 | Ahrenholz S, Cardarelli J, Dill P et al (2001) Final Report. Mortality patterns among workers at the Portsmouth gaseous diffusion plant Piketon, Ohio. NIOSH |
| 13.5 | 1.15 | 0.67–1.83 | Boice JD, Cohen SS, Mumma MT et al (2006) Mortality among radiation workers at Rocketdyne (Atomics International), 1948–1999. Radiat Res 166:98–115 |
| 3 | 1.16 | | |
| 7 | 1.22 | | |
| 25 | 0.77 | | |
| 75 | 4.12 | | |
| 13.5 | 0.85 | 0.55–1.26 | Zablotska LB, Ashmore JP, Howe GR (2004) Analysis of mortality among Canadian nuclear power industry workers after chronic low-dose exposure to ionizing radiation. Rad Res 161:633–641 |
| 10 | 0.51 | 0.18–1.43 | |
| 70 | 0.54 | 0.08–3.35 | |
| 25.7 | 0.85 | 0.54–1.28 | Howe GR, Zablotska LB, Fix JJ et al (2004) Analysis of the mortality experience amongst U.S. nuclear power industry workers after chronic low-dose exposure to ionizing radiation. Rad Res 162:517–526 |
| 25 | 0.60 | 0.20–1.78 | |
| 75 | 0.41 | 0.05–3.74 | |

**Table A12.4** (continued)

| Dose (mSv) | RR (SMR) | 90–95% CI | Reference |
|---|---|---|---|
| 109 | 2.14 | 1.07–3.83 | Rahu M, Rahu K, Auvinen A et al (2006) Cancer risk among Chernobyl cleanup workers in Estonia and Latvia, 1986–1998. Int J Cancer 119:162–168 |
| 13 | 0.94 | 0.61–1.44 | Schubauer-Berigan MK, Macievic GV, Utterback DF et al (2005) Final report. An epidemiologic study of mortality and radiation-related risk of cancer among workers at the Idaho National Engineering and Environmental Laboratory, a U.S. Department of Energy facility. NIOSH |
| 7 | 0.51 | 0.17–1.19 | Silver SR, Anderson-Mahoney P, Burphy J et al (2005) Mortality update for the Pantex weapons facility: final report health-related energy research branch. National Institute for Occupational Safety and Health, HHS (NIOSH) Publication No. 2005–124 |
| 17.9 | 1.56 | 0.98–2.37 | Rogel A, Carre N, Amoros E et al (2005) Mortality of workers exposed to ionizing radiation at the French National Electricity company. Am J Indust Med 47:72–82 |
| 117 | 1.42 | 0.56–2.98 | Band PR, Le ND, Fang R et al (1996) Cohort study of Air Canada pilots: mortality, cancer incidence, and leukemia risk. Am J Epidemiol 143:137–143 |
| 678 (lung) | 0.70 | 0.14–2.04 | Wu X, Jiang R (2006) Epidemiology investigate on mortality of uranium miner in Jiangxi Province. Abstracts of The Second Asian and Oceanic Congress for Radiation Protection, Beijing, China, p 354; Wu X (2006) Personal Communication |
| Uranium Miners | 0.95 | 0.71–1.25 | Darby SC, Whitley E, Howe GR et al (1995) Radon and cancers other than lung cancer in underground miners: a collaborative analysis of 11 studies. J Natl Cancer Inst 87:378–384 |
| 10 | 1.81 | – | McGeoghegan D, Binks K (2000) The mortality and cancer morbidity experience of workers at the Capenhurst uranium enrichment facility 1946–95. J Radiol Prot 20:381–401 |
| 19 | 1.00 | 0.57–1.77 | Atkinson WD, Law DV, Bromley KJ et al (2004) Mortality of employees of the United Kingdom Atomic Energy Authority, 1946–97. Occup Environ Med 61:577–585 |
| 5 | 1.04 | – | |
| 15 | 1.29 | – | |
| 35 | 0.91 | – | |
| 75 | 0.70 | – | |
| 130 | 0.91 | – | |

(*continued*)

**Table A12.4** (continued)

| Dose (mSv) | RR (SMR) | 90–95% CI | Reference |
|---|---|---|---|
| 18 | 0.8 | 0.2–2.9 | Ronckers CM, van Leeuwen FE, Hayes RB et al (2002) Cancer incidence after nasopharyngeal radium irradiation. Epidemiol 13:552–560 |
| 97 | 0.87 | – | Sponsler R, Cameron JR (2005) Nuclear shipyard worker study (1980–1988): a large cohort exposed to low-dose-rate gamma radiation. Int J Low Radiation 1:463–478 |
| 100 | 0.82 | 0.09–3.25 | Blettner M, Zeeb H, Langner I et al (2002) Mortality from cancer and other causes among airline cabin attendants in Germany, 1960–1997. Am J Epidemiol 156:556–565 |
| 32 | 0.79 (SIR) | 0.67–0.93 | Sont WN, Zielinski JM, Ashmore JP et al (2001) First analysis of cancer incidence and occupational radiation exposure based on the National Dose Registry of Canada. Am J Epidemiol 153:309–318 |
| 26.5 | 0.25 | 0.01–1.21 | Wiggs LD, Cox-DeVore CA, Voelz GL (1991) Mortality among a cohort of workers monitored for 210Po exposure: 1944–1972. Health Phys 61:71–76 |
| 21 | 0.68 | 0.43–1.02 | Wiggs LD, Cox-DeVore CA, Voelz GL (1991) Mortality among a cohort of workers monitored for 210Po exposure: 1944–1972. Health Phys 61:71–76 |
| 23.3 | 0.86 | – | Gilbert ES, Omohundro E, Buchanan JA et al (1993) Mortality of workers at the Hanford site: 1945–1986. Health Phys 64:577–590 |
| 5.8 | 0.71 | – | Wilkinson GS, Trieff N, Graham R et al (2000) Final report. Mortality among female nuclear weapons workers. NIOSH |
| 35 | 0.89 | 0.46–1.66 | Carpenter LM, Higgins CD, Douglas AJ et al (1998) Cancer mortality in relation to monitoring for radionuclide exposure in three UK nuclear industry workforces. Br J Cancer 78:1224–1232 |
| 9.3 | 0.83 | – | McGeoghegan D, Binks K (2000) The mortality and cancer morbidity experience of workers at the Springfields uranium production facility, 1946–95. J Radiol Prot 20:111–137 |
| 800 | 0.65 (SIR) | 0.07–2.35 | Ron E, Auvinen A, Alfandary E et al (1999) Cancer risk following radiotherapy for infertility or menstrual disorders. Int J Cancer 82:795–798 |

**Table A12.4** (continued)

| Dose (mSv) | RR (SMR) | 90–95% CI | Reference |
|---|---|---|---|
| 70 WLM | 1.89 | 0.76–3.89 | Tirmarche M, Raphalen A, Allin F et al (1993) Mortality of a cohort of French uranium miners exposed to relatively low radon concentrations. Br J Cancer 67:1090–1097 |
| 280 | 1.20 | 0.60–2.40 | Franklyn JA, Maisonneuve P, Sheppard M et al (1999) Cancer incidence and mortality after radioiodine treatment for hyperthyroidism: a opoulation-based cohort study. Lancet 353:2111–2115 |
| 60 | 0.77 M<br>0.92 F | 0.6–1.0<br>0.7–1.1 | Mohan AK, Hauptmann M, Freedman DM et al (2003) Cancer and other causes of mortality among radiologic technologists in the United States. Int J Cancer 103:259–267 |
| 77 | 1.85 (SIR) | 1.28–2.59 | Lindberg S, Karlsson P, Arvidsson B et al (1995) Cancer incidence after radiotherapy for skin haemangioma during infancy. Acta Oncol 34:735–740 |
| 35 | 0.95 | 0.75–1.16 | Sigurdson AJ, Doody MM, Rao RS et al (2003) Cancer incidence in the U.S. radiologic technologists health study, 1983–1998. Cancer 97:3080–3089 |
| 110 | 0.63 | 0.00–3.62 | Reynolds P, Cone J, Layefsky M et al (2002) Cancer incidence in California flight attendants (United States). Cancer Causes Control 13:317–324 |
| ~10 | 0.93 (M)<br>1.57 (F) | 0.71–1.19<br>0.88–2.59 | Telle-Lamberton M, Bergot D, Gagneau M et al (2004) Cancer mortality among French Atomic Energy Commission workers. Am J Indust Med 45:34–44 |
| 30<br>75<br>140 | 1.28<br>9.22<br>3.26 | –<br>–<br>– | Wiggs LD, Johnson ER, Cox-DeVore CA et al (1994) Mortality through 1990 among white male workers at the Los Alamos National Laboratory: considering exposures to plutonium and external ionizing radiation. Health Phys 67:577–588 |
| 48 | 1.57 | 0.84–2.64 | Dupree-Ellis E, Watkins J, Ingle JN et al (2000) External radiation exposure and mortality in a cohort of uranium processing workers. Am J Epidemiol 152:91–95 |
| 6.3 | 0.73 (M)<br>0.58 (F) | 0.57–0.91<br>0.37–0.87 | Ashmore JP, Krewski D, Ziellnski JM et al (1998) First analysis of mortality and occupational radiation exposure based on the National Dose Registry of Canada. Am J Epidemiol 148:564–574 |

*(continued)*

**Table A12.4** (continued)

| Dose (mSv) | RR (SMR) | 90–95% CI | Reference |
|---|---|---|---|
| 1,400 | 3.0 (SIR) | 1.3–5.9 | Shore RE, Moseson M, Harley N et al (2003) Tumors and other diseases following childhood X-ray treatment for ringworm of the scalp (*Tinea capitis*). Health Phys 85:404–408 |
| 30.5 | 0.90 | – | Muirhead CR, Goodill AA, Haylock RGE et al (1999) Occupational radiation exposure and mortality: second analysis of the National Registry of Radiation Workers. J Radiol Prot 19:3–26 |
| 29 | 0.67 (F)<br>0.94 (M) | 0.27–1.49<br>0.33–2.11 | Zeeb H, Blettner M, Langner I et al (2003) Mortality from cancer and other causes among airline cabin attendants in Europe: a collaborative cohort study in eight countries. Am J Epidemiol 158:35–46 |
| ~20 | 0.58 | 0.22–1.52 | Grayson JK (1996) Radiation exposure, socioeconomic status, and brain tumor risk in the US Air Force: a nested case-control study. Am J Epidemiol 143:480–486 |
| 260 | 1.06 | – | Thompson DE, Mabuchi K, Ron E et al (1994) Cancer incidence in atomic bomb survivors. Part II. Solid tumors, 1958–1987. Radiat Res 137:S17–S67 |
| 1,500 | 7.14 | – | Ron E, Modan B, Boice JD et al (1988) Tumors of the brain and nervous system after radiotherapy in childhood. N Engl J Med 319:1033–1039 |
| 1,400 | 5.71 | – | Albert RE, Shore RE, Harley N et al (1986) Follow-up studies of patients treated by x-ray epilation for tinea capitis. In: Burns FJ, Upton AC, Silini G (eds) Radiation carcinogenesis and DNA alterations. Plenum Press, New York, pp 1–25 |
| 70 | 1.43 | – | Karlsson P, Holmberg E, Lundell M et al (1998) Intracranial tumors after exposure to ionizing radiation during infancy. A pooled analysis of two Swedish cohorts of 28,008 infants with skin hemangioma. Radiat Res 150:357–364 |
| 678 | 0.70 | 0.14–2.04 | Xiao WU, Jiang R, Chang X et al (2006) Epidemiologist investigate on mortality of uranium miner in Jiangxi Province. Proceedings of The Second Asian and Oceanic Congress for Radiation Protection, October 9–13, 2006, Beijing, China, pp 1314–1318 |

**Table A12.5** Risk of thyroid cancer in epidemiological studies of populations exposed to ionizing radiation

| Dose (mSv) | RR (SMR) | 90–95% CI | Reference |
|---|---|---|---|
| 200 | 0.84 | 0.15–4.61 | Tao Z, Zha Y, Akiba S et al (2000) Cancer mortality in the high background radiation areas of Yangjiang, China during the period between 1979 and 1995. J Radiat Res 41(Suppl):31–41 |
| 20 | 0.62 | 0.02–3.42 | Pukkala E, Auvinen A, Wahlberg G (1995) Incidence of cancer among Finnish airline cabin attendants, 1967–92. BMJ 311:649–652 |
| 13 | 2.85 | 0.32–25.6 | Schubauer-Berigan MK, Macievic GV, Utterback DF et al (2005) Final report. An epidemiologic study of mortality and radiation-related risk of cancer among workers at the Idaho National Engineering and Environmental Laboratory, a U.S. Department of Energy facility. NIOSH |
| 48 | 2.6 (SIR) | 1.1–5.4 | Hwang SL, Guo HR, Hsieh WA et al (2006) Cancer risks in a population with prolonged low dose-rate γ-radiation exposure in radiocontaminated buildings, 1983–2002 Int J Radiat Biol 82:849–858 |
| 19 | 0.83 | 0.18–3.28 | Atkinson WD, Law DV, Bromley KJ et al (2004) Mortality of employees of the United Kingdom Atomic Energy Authority, 1946–97. Occup Environ Med 61:577–585 |
| 551<br>82 | 2.06<br>0.39 | –<br>– | Wang JX, Zhang LA, Li BX et al (2002) Cancer incidence and risk estimation among medical X-ray workers in China, 1950–1995. Health Phys 82:455–466 |
| 32 | 1.39 (SIR) | 1.20–1.61 | Sont WN, Zielinski JM, Ashmore JP et al (2001) First analysis of cancer incidence and occupational radiation exposure based on the National Dose Registry of Canada. Am J Epidemiol 153:309–318 |
| 5.8 | 1.14 | – | Wilkinson GS, Trieff N, Graham R et al (2000) Final report. Mortality among female nuclear weapons workers. NIOSH |
| 35 | 0.15 | 0.01–0.89 | Carpenter LM, Higgins CD, Douglas AJ et al (1998) Cancer mortality in relation to monitoring for radionuclide exposure in three UK nuclear industry workforces. Br J Cancer 78:1224–1232 |
| 145<br>316<br>450 | 0.5<br>1.3<br>2.1 | 0.05–4.0<br>0.2–11<br>0.2–18 | Lyon JL, Alder SC, Stone MB et al (2006) Thyroid disease associated with exposure to the Nevada nuclear weapons test site radiation. A reevaluation based on corrected dosimetry and examination data. Epidemiol 17:604–614 |

*(continued)*

**Table A12.5** (continued)

| Dose (mSv) | RR (SMR) | 90–95% CI | Reference |
|---|---|---|---|
| 940 | 0.91 | 0.64–1.26 | Dickman PW, Holm L-E, Lundell G et al (2003) Thyroid cancer risk after thyroid examination with 131I: a population-based cohort study in Sweden. Int J Cancer 106:580–587 |
| 400 | 1.1 (SIR) | 0.3–2.8 | Hamilton PM, Chiacchierini RP, Kaczmarek RG (1989) A follow-up study of persons who had iodine-131 and other diagnostic procedures during childhood. US Department of Health and Human Services, Public Health, Food and Drug Administration, Publication FDA 89–827 |
| 1,000 | 0.86 | 0.14–5.13 | Hahn K, Schnell-Inderst P, Grosche B et al (2001) Thyroid cancer after diagnostic administration of iodine-131 in childhood. Radiat Res 156:61–70 |
| 60 | 0.79 | 0.3–1.7 | Mohan AK, Hauptmann M, Freedman DM et al (2003) Cancer and other causes of mortality among radiologic technologists in the United States. Int J Cancer 103:259–267 |
| 120 | 1.88 | 1.05–3.09 | Lindberg, S, P Karlsson, B Arvidsson et al. 1995. Cancer incidence after radiotherapy for skin haemangioma during infancy. Acta Oncol 34:735–740 |
| 35 | 1.61 | 1.34–1.88 | Sigurdson AJ, Doody MM, Rao RS et al (2003) Cancer incidence in the U.S. radiologic technologists health study, 1983–1998. Cancer 97:3080–3089 |
| 110 | 0.25 | 0.00–1.43 | Reynolds P, Cone J, Layefsky M et al (2002) Cancer incidence in California flight attendants (United States). Cancer Causes Control 13:317–324 |
| ~10 | 0.71 | 0.24–1.62 | Telle-Lamberton M, Bergot D, Gagneau M et al (2004) Cancer mortality among French Atomic Energy Commission workers. Am J Indust Med 45:34–44 |
| 30.5 | 1.52 | – | Muirhead CR, Goodill AA, Haylock RGE et al (1999) Occupational radiation exposure and mortality: second analysis of the National Registry of Radiation Workers. J Radiol Prot 19:3–26 |
| 1,000 | 1.60 | 1.00–2.42 | Damber L, Johansson L, Johansson R et al (2002) Thyroid cancer after X-ray treatment of benign disorders of the cervical spine in adults. Acta Oncol 41:25–28 |

**Table A12.5** (continued)

| Dose (mSv) | RR (SMR) | 90–95% CI | Reference |
|---|---|---|---|
| 380 | 1.79 | 1.05–2.98 | Hall P, Mattsson A, Boice JD (1996) Thyroid cancer after diagnostic administration of iodine-131. Radiat Res 145:86–92 |
| 750 | 0.45 | 0.38–1.57 | |
| 1,250 | 1.55 | 1.09–2.26 | |
| 260 | 1.40 | – | Thompson DE, Mabuchi K, Ron E et al (1994) Cancer incidence in atomic bomb survivors. Part II. Solid tumors, 1958–1987. Radiat Res 137:S17–S67 |
| 8,200 | 40 | – | Shore RE (1992) Issues and epidemiological evidence regarding radiation-induced thyroid cancer. Radiat Res 131:98–111 |
| 100 | 1.43 | – | |
| 240 | 2.41 | – | |
| 2,900 | 14.4 | – | |
| 4,500 | 55.0 | – | |
| 100 | 4.02 | – | Ron E, Modan B, Preston D et al (1989) Thyroid neoplasia following low-dose radiation in childhood. Radiat Res 120:516–531 |
| 1,400 | 13.7 | – | Shore RE, Hildreth N, Dvoretsky P et al (1993) Thyroid cancer among persons given x-ray treatment in infancy for an enlarged thymus gland. Am J Epidemiol 137:1068–1080 |
| 12,500 | 57.5 | | Tucker MA, Morris Jones PH, Boice JD et al (1991) Therapeutic radiation at a young age is linked to secondary thyroid cancer. Cancer Res 51:2885–2888 |
| 260 | 2.67 | – | Lundell M, Hakulinen T, Holm L-E (1994) Thyroid cancer after radiotherapy for skin hemangioma in infancy. Radiat Res 140:334–339 |
| 600 | 2.80 | – | Schneider AB, Ron E, Lubin J et al (1993) Dose-response relationships for radiation-induced thyroid cancer and thyroid nodules: evidence for the prolonged effects of radiation on the thyroid. J Clin Endocrinol Metab 77:362–369 |
| 110 | 2.39 | – | Boice JD, Engholm G, Kleinerman RA et al (1988) Radiation dose and second cancer risk in patients treated for cancer of the cervix. Radiat Res 116:3–55 |
| 110 | 1.28 | – | Boice JD, Day NE, Andersen A et al (1985) Second cancers following radiation treatment for cervical cancer. An international collaboration among cancer registries. J Natl Cancer Inst 74:955–975 |
| 45,000 | 15 | – | Hancock SL, Cox RS, McDougall R (1991) Thyroid diseases after treatment of Hodgkin's disease. N Engl Med 325:599–605 |

## References


1. Jasworowski Z (2009) Radiation hormesis—a remedy for fear. BELLE Newsletter 15:14–20
2. Colombo M (1992) Hepatocellular carcinoma. J Hepatol 15:225–236
3. Sanders CL (1996) Prevention and therapy of cancer and other common diseases: alternative and traditional approaches. CD-ROM, Ð3000 pages in HTML and MS Word. Infomedix, Richland, WA
4. Fujiwara S, Kusumi S, Cologne J et al (2000) Prevalence of anti-hepatitis C virus antibody and chronic liver disease among atomic bomb survivors. Radiat Res 154:12–19
5. Cologne JB, Tokuoka S, Beebe GW et al (1999) Effects of radiation on incidence of primary liver cancer among atomic bomb survivors. Radiat Res 152:364–373
6. van Kaick G, Lorenz D, Muth H, Kaul A (1978) Malignancies in German Thorotrast patients and estimated tissue dose. Health Phys 35:127–136
7. da Silva Horta J, da Silva Horta ME, da Motta L et al (1978) Malignancies in Portuguese Thorotrast patients. Health Phys 35:137–151
8. Van Kaick G, Wesch H, Luhrs H et al (1991) Neoplastic diseases induced by chronic alpha irradiation. Epidemiological, biophysical and clinical results by the German Thoratrast study. J Radiat Res 32:20–33
9. Tokarskaya ZB, Zhuntova GV, Scott BR et al (2006) Influence of alpha and gamma radiations and non-radiation risk factors on the incidence of malignant liver tumors among Mayak workers. Health Phys 91:296–310
10. Hankey BF (1992) Brain and other nervous system. In: Niller BA, Gloeckler LA, Kosary CL, Edwards BK (eds) NIH Publ. 92–2789, Bethesda, MD
11. Perry JR, Ang LC, Bilbao JM, Muller PJ (1995) Clinicopathologic features of primary and postirradiation cerebral gliosarcoma. Cancer 75:2910–2918
12. Ron E, Modan B, Boice J et al (1988) Tumors of the brain and nervous system following radiotherapy in childhood. N Engl J Med 319:1033–1039
13. Shimizu Y, Kato H, Schull W (1990) Studies of the mortality of A-bomb survivors. 9. Mortality, 1950–1985: Part 2. Cancer mortality based on the recently revised doses (DS86). Radiat Res 121:120–141
14. Van Kaick G, Muth M, Kaul A et al (1986) Report on the German Thorotrast Study. Strahlentherapie 80(Suppl):114–118
15. Wilkinson GS, Tietjen GL, Wiggs LD et al (1987) Mortality among plutonium and other radiation workers at a plutonium weapons facility. Am J Epidemiol 125:231–250
16. Hadjimichael OC, Ostfeld AM, D'Atri DA (1983) Mortality and cancer incidence experience of employees in a nuclear fuels fabrication plant. J Occup Med 25:48–61
17. Sanders CL, Dagle GE, Mahaffey JA (1992) Incidence of brain tumors in rats exposed to an aerosol of $^{239}PuO_2$. Int J Radiat Biol 62:97–102
18. Lonser R, Walbridge S, Vortmeyer A et al (2002) Induction of glioblastoma multiforme in nonhuman primates after therapeutic doses of fractionated whole-brain radiation therapy. J Neurosurg 97:1378–1389
19. Ron E, Lubin JH, Shore RE et al (1995) Thyroid cancer after exposure to external irradiation: a pooled analysis of seven studies. Radiat Res 141:259–277
20. Thompson DE, Mabuchi K, Ron E et al (1994) Cancer incidence in atomic bomb survivors. Part II. Solid tumors 1958–1987. Radiat Res 137:S17–S67
21. Tubiana M (2000) Radiation risks in perspective: radiation-induced cancer among cancer risks. Radiat Environ Biophys 39:3–16
22. National Council on Radiation Protection (1985) Induction of thyroid cancer by ionizing radiation. NCRP Report No. 80, Chapter 4. Human experience after exposure to Iodine-131
23. Brown RA (1997) Bomb fallout and thyroid cancer: statistical sheep in real wolves' clothing. http://www.srv.net/Ðrussb/thyroid/index.html

24. National Cancer Institute (1997) NCI releases results of nationwide study of radioactive fallout from nuclear tests, Bethseda, MD, August 1, 1997. http://rex.nci.nih.gov/INTRFCE_GIFS/MASSMED_INTR_DOC.htm
25. Cardis E, Kesminienne A, Ivanov V et al (2005) Risk of thyroid cancer after exposure to I-131 in childhood. J Natl Cancer Inst 97:724–732
26. Scott BR (2006) Re: Risk of thyroid cancer after exposure to (131)I in childhood. J Natl Cancer Inst 98:561
27. Hancock SL, McDougall IR, Constine LS (1995) Thyroid abnormalities after therapeutic external radiation. Int J Radiat Oncol Biol Phys 31:1165–1170
28. Modan B, Bidatz D, Mart H (1974) Radiation-induced head and neck tumors. Lancet 1:277–279
29. Ron E, Modan B (1980) Benign and malignant thyroid neoplasms after childhood irradiation for tinea capitis. J Natl Cancer Inst 65:7–11
30. Moosa M, Mazzaferri EL (1997) Occult thyroid carcinoma. Cancer J 10:180–188
31. Furmanchuk AW, Roussak N, Ruchti C (1993) Occult thyroid carcinomas in the region of Minsk, Belarus. An autopsy study of 215 patients. Histopathology 23:319–325
32. Harach HR, Franssila KO, Wasenius VM (1985) Occult papillary carcinoma of the thyroid- A "normal" finding in Finland. A systematic study. Cancer 56:531–538
33. Fornalski KW, Dobrzynski L (2010) The healthy worker effect and nuclear industry workers. Dose-Response (in press)
34. Ron E, Lubin J, Schneider AB (1992) Thyroid cancer incidence. Nature 360:113
35. Bennett B, Repacholi M, Carr Z (eds) (2006) Health effects of the Chernobyl accident and special health care programmes. World Health Organization, Geneva, p 5
36. Hatch M, Ron E, Bouville A et al The Chernobyl disaster: cancer following the accident at the Chernobyl nuclear power plant. Epidemiol Reviews 2005: 27:56–66
37. Chernobyl Forum (IAEA, WHO, UNDP, UNEP, UN-OCHA, UNSCEAR, World Bank) (2005) Chernobyl's Legacy: Health, Environmental and Socio-Economic Impacts. The work is in three volumes and 600 pages by more than 100 scientists
38. Hall P, Mattsson A, Boice JD (1996) Thyroid cancer after diagnostic administration of iodine-131. Radiat res 145:86–92
39. Holm LE, Hall P, Wiklud K et al (1991) Cancer risk after iodine-131 therapy for hyperthyroidism. J Natl Cancer Inst 83:1072–1077
40. Holm LE, Wiklud K, Lundell G et al (1988) Thyroid cancer after diagnostic doses of iodine-131: A retrospective cohort study. J Natl Cancer Inst 80:1133–1138
41. Yalow RS (1995) Radiation and public perception. In: Young JP, Yalow RS (eds) Radiation and public perception, benefits and risks, American Chemical Society, Washingston, DC, pp 13–22
42. de Vathaire F, Schumberger M, Delisle MJ et al (1997) Leukaemias and cancers following I-131 administration for thyroid cancer. Br J Cancer 75:734–739
43. Franklyn JA, Maisonneuve P, Sheppard M et al (1999) Cancer incidence and mortality after radioiodine treatment for hyperthyroidism: a population-based cohort study. Lancet 353:2111–2115
44. USDHHS (U.S. Department of Health and Human Services, Center for Disease Control and Prevention, National Center for Environmental Health) (2002) The Hanford Thyroid Disease Study. http://www.cdc.gov/nceh/radiation/hanford/htdsweb/index.htm

# 13 Lifespan, Birth Defects, and Experimental Cancer

Increased ionizing radiation in radiation-deficient environments provides an increased lifespan and abundant health

(TD Luckey)

## 13.1 Lifespan

The Radiation Effects Research Foundation (RERF), which is a cooperative Japan–U.S. study of Japanese A-bomb survivors, has several research functions. Among them are life span, birth defect, and *in utero* studies (Table 13.1).

Effects of aging can be slowed or prevented by lifestyle changes, nutritional strategies and steady, continuous exercise sufficient to make muscle cells need increased amounts of oxygen. Feeding rodents *ad libitum* nutritionally rich, high-calorie diets results in poor survival compared with the beneficial effects of simple caloric restriction. The beneficial effects of caloric restriction on lifespan have been documented in invertebrates, rodents, and other vertebrates such as fish, birds, other mammals, and humans. A 30–40% caloric restriction in rodents lowers the incidence and/or delays the onset of most spontaneous tumors, reduces the severity and/or onset of many spontaneous degenerative diseases, and extends average and maximal lifespan [1–4].

Increased lifespan has been observed in many lifespan studies of irradiated mice, rats and dogs (Table 13.2) [5–10]. Male LAF1 mice exposed for duration of life to ~1 mSv/day significantly increased survival over unexposed controls [5]. Whole-body X-irradiation of mice with 5 cGy significantly increased survival [11]. A single dose of 500 or 1,000 mSv 250 Kvp X-rays to 7- or 21-day-old mice increased lifespan [12]. Whole-body $\gamma$-irradiated C57BL/6 mice at doses of 70 and 140 mSv/year experienced significant ($p < 0.01$) prolongation of lifespan [6]. Successive generations of rats exposed continuously at 20 mSv/day to $^{60}$Co $\gamma$-rays had significantly longer lifespan, more robust reproduction and fewer tumors than unexposed controls [14]. Diabetic mice receiving 0.70 mGy/h of continuous $\gamma$-irradiation (~6 Sv/year) lived significantly longer than unirradiated diabetic mice.

C. L. L. Sanders, *Radiation Hormesis and the Linear-No-Threshold Assumption*, DOI: 10.1007/978-3-642-03720-7_13, © Springer Verlag Berlin Heidelberg 2010

**Table 13.1** Research programs sponsored by RERF

| Study | Population size | Study period |
|---|---|---|
| Life span | 120,000 | >1949 |
| Adult health | 20,000 | >1957 |
| *In utero* | 3,600 | >1945 |
| Mortality | 77,000 | >1947 |
| Birth defects | 77,000 | 1948–1954 |
| Cancer incidence | 77,000 | >1957 |
| Chromosome aberrations | 16,000 | 1967–1984 |
| Molecular genetics | 1,500 | >1984 |

**Table 13.2** Early studies of low-dose, low LET irradiation in rodents that demonstrated a prolonged lifespan

| Year | Author | Radiation | Animal |
|---|---|---|---|
| 1950 | Lorenz | X-rays | Mice |
| 1955 | Maisin | X-rays | Mice |
| 1956 | Sacher | X-rays | Mice |
| 1957 | Carlson | γ-rays | Rats |
| 1958 | Lindop | X-rays | Mice |
| 1960 | Gowen | X-rays | Mice |
| 1962 | Sacher | γ-rays | Rats |

Chronic low-dose-rate gamma irradiation of MRL-*lpr/lpr* mice (carry a deletion in the apoptosis-regulating *Fas* gene that markedly shortens lifespan due to lymphadenopathy, splenomegaly, and severe autoimmune diseases) at 0.35 or 1.2 mGy/h for 5 weeks markedly prolonged lifespan due to immunological activation and stimulation of apoptosis. The 50% survival time for untreated mice was 134 days, which increased to 502 days in mice given life-long whole body gamma irradiation at 1.2 mGy/h (Fig. 2.13) [15, 16].

A U-shaped lifespan pattern was found in A-bomb survivors, with increased lifespan at low doses (Table 13.3, Fig. 13.1) [17–19]. Accelerated aging was noted by the early 1970s in Japanese A-bomb survivors who received radiation doses >1 Gy [20]. However, mortality rates in 120,321 atomic bomb survivors was not increased at doses <490 mSv [21]. Lifespan was prolonged in Japanese A-bomb survivors who received whole-body doses of <149 mGy [17, 22]. The lifespan of A-bomb survivors was greater than in control groups [17].

Studies of early radiologists showed lifespan shortening. Later studies of radiologists did not show this effect, presumably because of better radiation protection practices, but showed a significant decrease in all cause mortality [23–25]. Participants in the United States Transuranium and Uranium Registries (USTUR) significantly exceeded "life table" longevity expectation by an average of 10.4 years [26]. Both the U.S. and British radium dial painters exhibited an increased lifespan [27]. Nuclear workers exhibit highly significant results in most disease categories. The mean decreases in all cause mortality and all cancer mortality were 38 and 26%, respectively, in nuclear workers employed in 154 facilities in 15 countries [28], which equates to a significant lifespan prolongation.

**Table 13.3** Relative risk for death from all causes and from noncancerous diseases among Nagasaki atomic bomb survivors; 80% of deaths caused by noncancer causes [17]

| Dose-range (Gy) | Relative risk | |
|---|---|---|
| | All causes | Noncancerous diseases |
| 0 | 1.00 | 1.00 |
| 0.005–0.05 | 1.02 | 1.01 |
| 0.06–0.09 | 0.96 | 0.93[a] |
| 0.10–0.19 | 0.96 | 0.92[a] |
| 0.20–0.49 | 0.97 | 0.93[a] |
| 0.50–0.99 | 1.04 | 0.97 |
| 1.00–1.99 | 1.07[a] | 0.92 |
| 2.00–6.00 | 1.42[a] | 1.15[a] |

[a] $p<0.05$

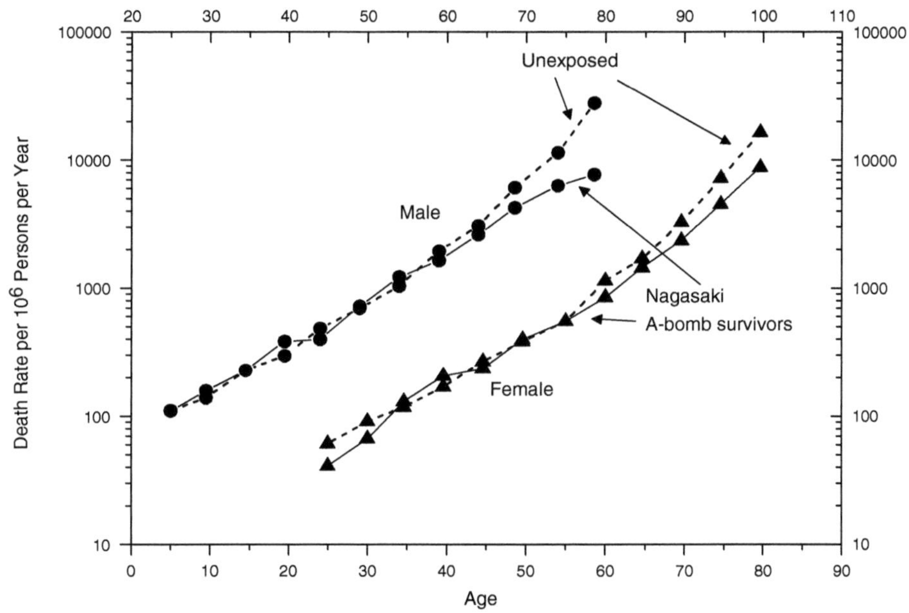


**Fig. 13.1** Mortality in male and female Japanese A-bomb survivors and comparable unexposed controls

## 13.2 Birth Defects

Radiation does not produce new, unique mutations but increases mutation frequency above spontaneously frequency. Information on the hereditary effects of radiation comes largely from animal experiments. No epidemiological study in humans has demonstrated significant genetic effects in offspring of irradiated parents [29]. No harmful genetic effects have

**Table 13.4** Effect of a 2 cGy priming dose in mice on teratogenesis from 2 Gy received 4 h later [34]

| Percentage of incidence | Control | 2 cGy | 2 Gy | 2 cGy + 2 Gy |
|---|---|---|---|---|
| Normal | 96.6 | 94.1 | 12.2 | 23.5 |
| Late mortality | 2.9 | 5.1 | 12.2 | 9.2 |
| Malformations | 0.5 | 0.8 | 75.5 | 67.3 |

been found in children of A-bomb survivors [30]. About 3,600 Japanese A-bomb survivor women were pregnant at the time of A-bomb explosions. Yet, no evidence of an increased incidence of birth defects was found. The threshold dose from X-ray exposure for lethality, mental retardation and growth retardation exposure of the conceptus was several hundred mSv [31].

Fear of radiation hazards were so great in countries surrounding the Chernobyl accident site that an estimated 100,000 women had voluntary abortions performed due to a fear of teratogenic effects [32]. In fact, women who received high-radiation doses experienced less birth defects than those exposed to low doses (Fig. 4.2) [33]. The increasing slope for birth defects in both dose groups was due to an enhanced screening effort following the Chernobyl accident.

Priming mice with a whole-body dose of 2 cGy (667 mGy/min) on day 9.5 of gestation significantly decreased radiation-induced teratogenesis from a dose of 2 Gy (1.04 Gy/min) given after a 4 h interval (Fig. 4.2) [34]. The effective *in vivo* hormetic, whole-body, priming dose in mice that reduced birth defect formation was 15–60 cGy [35].

Studies of Japanese A-bomb survivors exposed *in utero* did not suggest a prenatal carcinogenic effect, either in leukemia or solid tumor formation. Similarly, animal experiments have failed to find *in utero* radiation sensitivity to tumor formation. Studies of prenatal X-ray exposure and cancer in children failed to show a significant increased risk following prenatal radiation exposure.

## 13.3 Experimental Cancer

Evidence for both reduced tumor incidence and increased lifespan were frequently observed in studies where pathological examinations were carried out. Analyses of experimental animal data (85,000 exposed animals and 45,000 controls) showed frequent significant evidence of both reduced tumor incidence and increased lifespan from low dose radiation exposure [7]. A significantly ($p<0.01$) decreased incidence of cancer from expected spontaneous rates was observed in more than 30 published studies in rodents [36]. A reduction in cancer risk was observed in 40–61% of animal carcinogenesis studies involving exposures to neutrons, gamma rays, and alpha particles. Significant reductions of cancer were observed in mice exposed to 100 and 250 mGy of gamma irradiation [7]. The high frequency of reduced carcinogenic response coupled with increased lifespan in animals receiving low-dose, low-LET radiation places in question the validity of the LNT assumption.

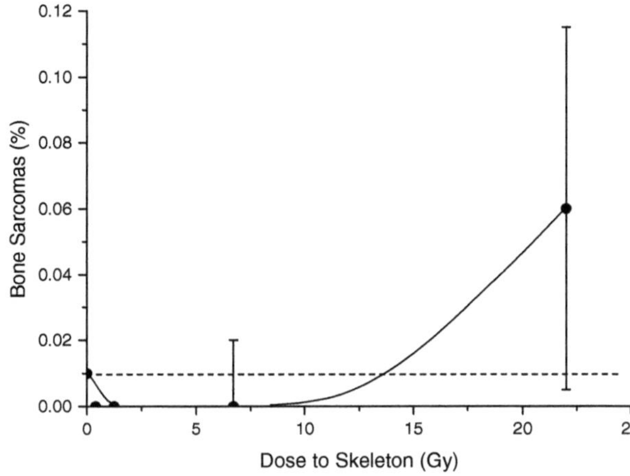


**Fig 13.2** Bone sarcoma incidence in beagle dogs given $^{90}Sr$

Most animal dose–response relationships for cancer formation are not linear, but are linear-quadratic, quadratic, and with a threshold and/or a hormetic effect [7, 37, 38]. Studies in experimental animals, mostly in rats, mice, and dogs, have failed to demonstrate a statistically significant carcinogenic risk for doses <100 mSv. With few exceptions, no significantly increased tumors are not found at doses <500 mSv for low-LET radiation [39]. Thresholds have been observed for lung tumors in rats and for bone tumors in dogs following exposure to α-radiation [40–45]. The incidence of bone sarcomas in dogs administered $^{90}Sr$ was less than expected (or zero) up to skeletal doses of about 7 Gy; increase sarcomas were seen only at a dose of 24 Gy (Fig 13.2). A reduction in tumors was observed following exposure to low-dose radiation in 40% of studies where unexposed control animals have a high spontaneous incidence of a specific tumor type [7]. Rhesus monkeys receiving 0.25–2.8 Gy from 32 to 2300 MeV protons had a tumor incidence of 3.5% (4/115) while the tumor incidence in unirradiated monkeys was 8.8% (5/57) [46, 47].

A reduced formation of mammary carcinomas was seen in rats after exposure to neutrons (0.05 Gy) or X-rays (0.25) and for pulmonary adenomas in mice after exposure to γ rays (0.25 Gy) [48, 49]. Mortality was significantly reduced in beagle dogs following injection of 0.0104 uCi/kg $^{239}Pu$ [50].

A whole-body priming dose of 0.1 Gy (8 mGy/min) given 24 h before a dose of 1.0 Gy significantly reduced lost lifespan in mice from acute myeloid leukemia as compared with mice receiving only 1.0 Gy [51]. Continuous whole-body X-irradiation of C57BL/6 mice at 1.2 mGy/h, for a total dose of 7.2 Gy more than 258 days, caused no thymic lymphomas to form; continued exposure up to 450 days (total dose of 12.6 Gy) also produced no tumors. The mice also had a greater body weight that unirradiated controls. A 90% incidence of thymic lymphoma formation was caused by acute fractionated exposures (4 × 1.8 Gy = 7.2 Gy); this level was decreased to 43% by additional exposure to 1.2 mGy/h for 450 days (Fig 13.3) [52]. Continuous lifespan exposure of SJL mice to γ-rays (10 cGy/year) increased lifespan, NK cell numbers in the spleen, but did not change the incidence of spontaneous B-cell lymphoma [53].

Dose-dependent increases in lung tumors have been found in rats and dogs exposed to a variety of α-emitting radionuclide aerosols, including $^{238}PuO_2$, $^{238}Pu(NO_3)_4$, $^{238}PuCl_4$,

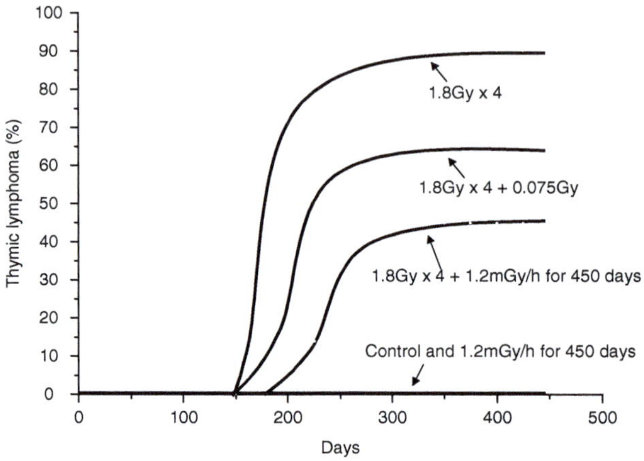


**Fig. 13.3** Suppression of acute X-irradiation induced thymic lymphoma induction in C57BL/6 mice by continuous low-dose X-irradiation. An acute dose of 1.8 Gy was given four times every 7 days for a total dose of 7.2 Gy (top curve); same as previous but first with a dose of 75 mGy (second from top curve); same as first curve but with continuous g-irradiation at 1.2 mGy/h for 450 days from 137Cs. The nonirradiated controls and mice given only continuous 1.2 mGy/h did not develop thymic lymphoma [52]

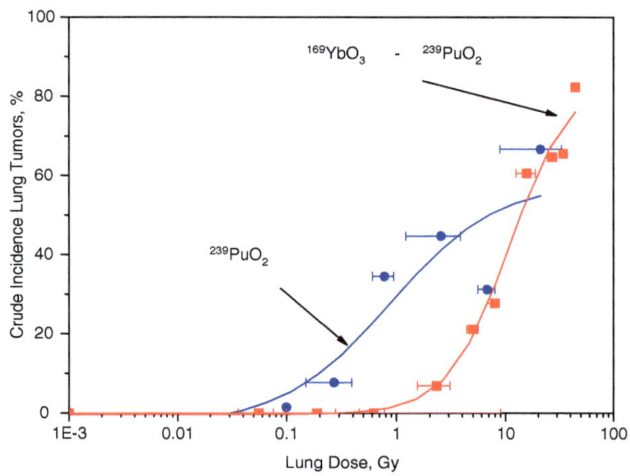


**Fig. 13.4** Frequency of lung tumors in 4,078 female Wistar rats following inhalation of $^{239}PuO_2$ or $^{169}YbO_3$–$^{239}PuO_2$ [44, 56]

$^{239}PuO_2$, $^{239}PuO_2$, $^{239}Pu(NO_3)_4$, $^{241}AmO_2$, and $^{244}CmO_2$ given by inhalation or intratracheal instillation to the lung of rats [54]. The RBE for α-emitting radionuclides was ~10 compared with β-emitting radionuclides [55].

Two nearly identical lifespan studies were carried out in the same laboratory with 70-day-old female Wistar rats exposed to submicron-sized, insoluble aerosols of high-fired $^{239}PuO_2$ particles. The first study [56] was with 936 rats exposed to $^{239}PuO_2$, and the second study was with 3,142 rats exposed to $^{169}Yb_2O_3$–$^{239}PuO_2$ [44, 57, 58]. The only difference between the two studies was that rats in the second study received a few mGy cumulative γ-ray doses from $^{169}Yb$ (Fig 13.4).

Lung doses were determined in the second study by whole body counting in individual rats using $^{169}Yb$ tagged to $^{239}PuO_2$ particles. $^{169}Yb$ emits 175 keV gamma rays. The effective half-life for $^{169}Yb$ in the lung was 14 days, with nearly all the dose being delivered in the first 3 months after exposure [58]. The γ-ray doses to the lung from $^{169}Yb$ following inhalation of $^{169}Yb$–$^{239}PuO_2$ ranged from 0.9 to 1.8, 1.3 to 5.0, and 0.8 to 7.0 mGy with lung doses of 56–190, 620, and >2,000 mGy, respectively, from $^{239}Pu$ derived 5.14 meV α-particles. About 80% of inhaled $^{239}Pu$ was cleared from the deep lung with a half-life of 79 days following a deep lung deposition of 0.4 kBq. The radiation dose-to-lung from $^{169}Yb$ was, at the most, 80 times less than the α-dose-to-lung from $^{239}Pu$ [58].

The second study contained 1,052 sham controls and 1,389 rats with mean lung doses of 0.06 Gy. Alpha lung doses delivered from $^{239}Pu$ ranged from 0.02 to 50 Gy. The lung tumor, threshold α-particle dose for $^{239}PuO_2$ was 0.05 Gy for the $^{239}PuO_2$-only first study (Table 13.5), while the lung tumor, threshold α-dose from $^{169}Yb_2O_3$–$^{239}PuO_2$ was 1.5 Gy. In the second study, no malignant lung tumors were found in 1,877 Pu-exposed rats at a lung dose <1.5 Gy (Table 13.6, Fig. 13.4). The threshold dose for the first appearance of squamous cell carcinoma was 1.5, 3.1 Gy for adenocarcinoma, 4.1 Gy for hemangiosarcoma, and 9 Gy for fibrosarcoma [44, 57]. A similar lung tumor response was seen in Mayak plutonium workers who also received workplace chronic γ-irradiation from $^{60}Co$ as

**Table 13.5** Frequency of lung tumors in female Wistar rats following inhalation of $^{239}PuO_2$ [56]. The lowest dose for a lung tumor in exposed rats was about 0.05 Gy

| Number rats | Lung dose (Gy) | Lung tumors (%) | Relative risk |
|---|---|---|---|
| 656 | 0 | 0.15 | 1.0 |
| 131 | <0.10 | 1.5 | 10 |
| 51 | 0.27 ± 0.12 | 7.8 | 52 |
| 26 | 0.78 ± 0.17 | 34.6 | 230 |
| 38 | 2.55 ± 1.32 | 44.7 | 298 |
| 16 | 6.80 ± 1.20 | 31.3 | 209 |
| 18 | 21.0 ± 12.1 | 66.7 | 445 |

**Table 13.6** Frequency of lung tumors in female Wistar rats following inhalation of $^{169}Yb_2O_3$–$^{239}PuO_2$ [44]. The lowest dose for a lung tumor in exposed rats was 1.5 Gy. The gamma ray dose to the lung from $^{169}Yb$ ranged from 1.6 to 7.0 mGy

| Number rats | Lung dose (Gy) | Lung tumors (%) | Relative risk |
|---|---|---|---|
| 1,052 | 0 | 0.095 | 1.0 |
| 1,389 | 0.056 ± 0.020 | 0 | – |
| 343 | 0.19 ± 0.09 | 0 | – |
| 145 | 0.62 ± 0.16 | 0 | – |
| 58 | 2.32 ± 0.77 | 6.9 | 73 |
| 38 | 5.03 ± 0.60 | 21.2 | 220 |
| 18 | 7.99 ± 0.67 | 27.8 | 290 |
| 33 | 15.7 ± 3.1 | 60.6 | 640 |
| 17 | 27.1 ± 2.7 | 64.7 | 680 |
| 32 | 34.5 ± 2.7 | 65.6 | 690 |
| 17 | 44.4 ± 3.1 | 82.3 | 870 |

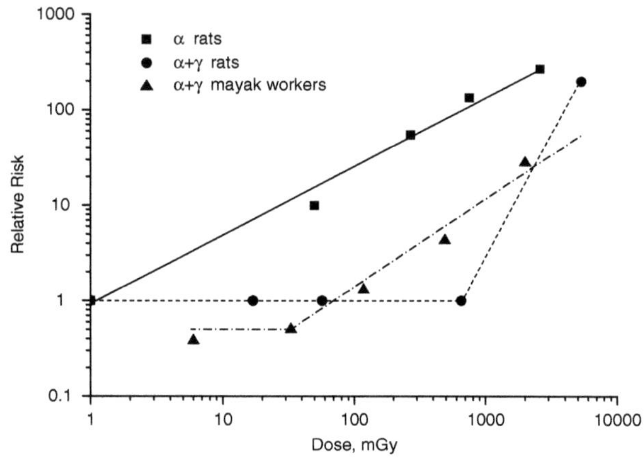


**Fig. 13.5** Relative risk of lung cancer in rats exposed to $^{169}$Yb γ-rays and plutonium and to Mayak nuclear workers exposed to γ-rays and plutonium [43, 59]

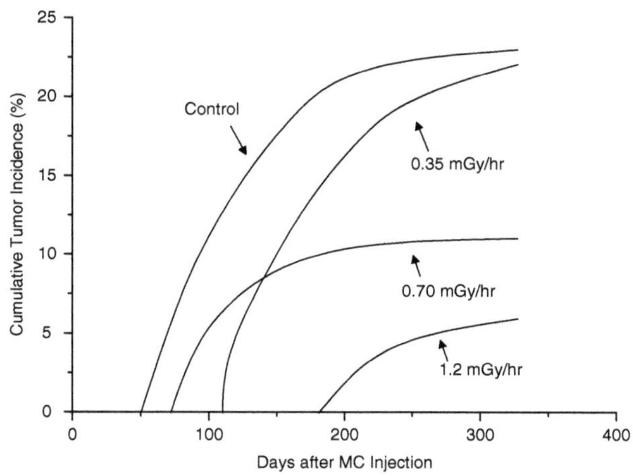


**Fig. 13.6** Suppression of methylcholanthrene induced skin tumors by low dose rate irradiation [60]

was observed in rats (Fig. 13.5) [43, 59]. These studies show that a small dose of low-LET radiation can prevent the formation of α-particle induced lung cancer in rats and humans.

Ionizing radiation is a relatively weak carcinogen and mutagen in animals as compared to chemical carcinogens, such as the polycyclic aromatic hydrocarbons. Low doses of low-LET radiation have been shown to protect animals from chemical carcinogenesis. A significant suppression of 20-methylcholanthrene carcinogenesis in mice was found in ICR female mice given γ-irradiation at 1 mGy/h (Fig. 13.6) [60]. A threshold dose-rate effect was observed for tumor formation in mouse skin by localized β-irradiation [61]. Exposure of the skin of mice to 50 cGy β-irradiation, 24 h before treatment with the chemical carcinogen, methyl-nitro-nitroso guanidine, reduced papilloma frequency by about fivefold [62]. Beta irradiation from tritium administered to mice in drinking water protected against formation of thymic lymphoma up to a dose-rate of 0.9 mGy/day [63]. Inhaled radon was much more effective in inducing lung carcinoma in rats when given before cigarette smoke. Radon given after cigarette smoke caused fewer lung carcinomas than when only radon

**Table 13.7** Lung carcinoma formation in rats exposed to radon and tobacco smoke [13]

| Study group | Percentage of lung carcinoma |
|---|---|
| Tobacco smoke only[a] | 0 |
| Radon only[b] | 22 |
| Radon[b] before tobacco smoke[a] | 78 |
| Tobacco smoke[1] before radon[b] | 16 |

[a]350 h
[b]1,000 WLM

was given (Table 13.7) [13]. Low LET irradiation from radon daughters may have induced apoptosis of smoke-induced transformed pulmonary cells.

The radioadaptive response at low doses is in contrast to damage at high doses, precluding a linear relationship between radiation dose and cancer incidence along the entire dose–response curve of experimental animal studies. Tumor thresholds are commonly seen in experimental studies of radiation carcinogenesis [7, 38, 48, 49, 64, 65]. Low-dose radiation stimulates many cellular functions that protects DNA from damage or from cells being expressed as cancer. The response includes removal of genomically damaged cells by apoptosis [36, 66–68]. The results of these inducible benefits explains the reduced carcinogenic effects in experimental animal studies receiving low-dose, low-LET radiation [69, 70].

## References


1. Weindruch R, Walford RL (1988) The retardation of aging and disease by dietary restriction. Charles C. Thomas, Springfield, IL
2. McCay CM, Crowell MF, Maynard LA (1935) The effect of retarded growth upon the length of the life span and upon the ultimate body size. J Nutr 10:63–79
3. Yu BP, Masoro J, McMahan CA (1985) Nutritional influences on aging of Fischer-344 rats. 1. Physical, metabolic and longevity characteristics. J Gerontol 40:657–670
4. Roe FJC (1990) 1200-Rat Biosure study: design and overview of results. In:Fishbein L (ed) Biological effects of dietary restriction, ILSI Monographs. Springer, NY, pp 287–304
5. Lorenz E, Hollcroft W, Miller E et al (1955) Long-term effects of acute and chronic irradiation of mice. I. Survival and tumor incidence following chronic irradiation of 0.11 r per day. J Natl Cancer Inst 15:1049–1058
6. Caratero A, Courtade M, Bonnet L et al (1998) Effect of a continuous gamma irradiation at a very low dose on the life span of mice. Gerontology 44:272–276
7. Duport P (2003) A database of cancer induction by low-dose radiation in mammals: overview and initial observations. Int J Low Radiat 1:120–131
8. Spalding JF, Thomas RG, Tietjen GI (1982) Life span of C57 mice as influenced by radiation dose, dose rate, and age at exposure. National Laboratory, U.S. Government Printing Office, Los Alamos
9. Maisin JR, Gerber5 GB, Vankerdom J, Wambersie A (1996) Survival and idseases in C57BL mice exposed to X rays or 3.1 MeV neutrons at an age of 7 or 21 days. Radiat Res 146:453–460
10. Carlson LD, Scheyer WJ, Jackson BH (1957) The combined effect of ionizing radiation and low temperature on the metabolism, longevity, and soft tissue of the rat. Radiat Res 7:190–197

11. Yonezawa M, Misonoh J, Hosokawa Y (1996) Two types of X-ray-induced radioresistance in mice: presence of 4 dose ranges with distinct biological effects. Mutat Res 358:237–243
12. Maisin J, Gerber G, Vankerkom J, Wambersie A (1996) Survival and diseases in C57BL mice exposed to X rays or 3.1 MeV neutrons at an age of 7 or 21days. Radiat Res 146:453–460
13. Monchaux G, Morlier JP, Morin M et al (1994) Carcinogenic and cocarcinogenic effects of radon and radon daughters in rats. Environ Health Perspect 102:64–73
14. Brown SO, Krise GM, Pace HB (1963) Continuous low-dose radiation effects on successive litters of the albino rat. Radiat Res 19:270–627
15. Ina Y, Sakai K (2004) Prolongation of life span associated with immunological modification by chronic low-dose-rate irradiation in MRL-*lpr/lpr* mice. Radiat Res 161:168–173
16. Ina Y, Sakai K (2005) Further study of prolongation of life span associated with immunological modification by chronic low-dose-rate irradiation in MRL-*lpr/lpr* mice: effects of whole-life irradiation. Radiat Res 163:418–423
17. Mine M, Okumura Y, Ichimaru M et al (1990) Apparently beneficial effect of low to intermediate doses of A-bomb radiation on human lifespan. Int J Radiat Biol 58:1035–1043
18. Kondo S (1993) Health effects of low-level radiation. Kinki University Press, Japan
19. Okumura Y, Mine M (1997) Effects of low doses of A-bomb radiation on human lifespan. Low doses of ionizing radiation: biological effects and regulatory control contributed papers. Seville: IAEA-TECDOC-976, IAEA-CN-67/129, International Atomic Energy Agency, Austria, pp 414–416
20. Anderson RE, Key CR, Yamamoto T, Thorslund T (1974) Aging in Hiroshima and Nagasaki atomic bomb survivors. Am J Pathol 75:1–12
21. Cologne J, Preston DL (2000) Longevity of atomic bomb survivors. Lancet 356:303–307
22. Okajima S, Mine M, Nakamura T (1985) Mortality of registered A-bomb survivors in Nagasaki, Japan, 1970–1984. Radiat Res 103:419–431
23. Berrington A, Darby SC, Weiss HA, Doll R (2001) 100 years of observation on British radiologists: mortality from cancer and other causes 1897–1997. Br J Radiol 74:507–519
24. Doody MM, Mandel JS, Lubin JH, Boice JD (1998) Mortality among United States radiologic technologists, 1926–90. Cancer Causes Control 9:67–75
25. Cameron JR (2002) Radiation increased the longevity of British radiologists. Br J Radiol 75:637–640
26. Fallahian N, Brey R, Watson C, James A (2007) Does exposure to plutonium affect workers' longevity? Health Phys 93:S11
27. Baverstock KF, Papworthy D (1987) The UK radium luminizer survey. Br J Radiol, Supplemental BIR Report 21, pp. 71–76
28. Vrijheid M, Cardis E, Blettner M et al (2007) The 15-country collaborative study of cancer risk among radiation workers in the nuclear industry: design, epidemiological methods and descriptive results. Radiat Res 167:361–379
29. UNSCEAR (2000) Sources and effects of ionizing radiation. United Nations Scientific Committee on the Effects of Atomic Radiation 2000 Report to the General Assembly, with Annexes. Volume II: Effects. No. E.00.IX.4. United Nations, New York, NY
30. Schull WJ (1998) The genetic effects of radiation: consequences for unborn life. Nucl Eur Worldscan (3–4):35–37
31. Damilakis J (2004) Pregnancy and diagnostic X-rays. Eur Radiol Syllabus 14:33–39
32. Cigna AA, Durante M (2006) Radiation risks in normal and emergency situations. Springer, Dordrecht, pp 49–67
33. Chernobyl Forum (IAEA, WHO, UNDP, UNEP, UN-OCHA, UNSCEAR, World Bank) (2005) Chernobyl's legacy: health, environmental and socio-economic impacts (The work is in three volumes and 600 pages by more than 100 scientists)
34. Okazaki R, Ootsuyama A, Norimura T (2005) Radioadaptive response for protection against radiation-induced teratogenesis. Radiat Res 163:266–270

35. Wang B, Ohyama H, Shang Y et al (2004) Adaptive response in embryogenesis: V. Existence of two efficient dose-rate ranges for 0.3 Gy of priming irradiation to adapt mouse fetuses. Radiat Res 161:264–272
36. Luckey TD (1991) Radiation hormesis. CRC, Boca Raton, FL
37. Tanooka H (2001) Threshold dose-response in radiation carcinogenesis: an approach from chronic beta-irradiation experiments and a review of non-tumour doses. Int J Radiat Biol 77:541–551
38. Upton AC (1989) The question of threshold for radiation and chemical carcinogenesis. Cancer Invest 7:267–276
39. Tubiana M, Aurengo A (2006) Dose-effect relationship and estimation of the carcinogenic effects of low doses of ionizing radiation: the Joint Report of the Academie des Sciences (Paris) and of the Academie Nationale de Medecine. Int J Low Radiat 2:1–19
40. Ogiso Y, Yamada Y, Iida H et al (1998) Differential dose responses of pulmonary tumor types in the rat after inhalation of plutonium dioxide aerosols. J Radiat Res 39:61–72
41. Kuschner M, Laskin S, Nelson N etal (1958) Radiation induced bronchogenic carcinoma in rats. Am J Pathol 34:554
42. Sanders CL, Mahaffey JA (1979) Inhalation toxicology of transuranics in rodents. In: Biological implications of radionuclides released from nuclear industries. IAEA, Vienna, Austria, p 89
43. Scott BR (2007) Low-dose radiation-induced protective process and implications for risk assessment, cancer prevention, and cancer therapy. Dose Response 5:131–141
44. Sanders CL, Lauhala KE, McDonald KE (1993) Lifespan studies in rats exposed to $^{239}PuO_2$ aerosol. III. Survival and lung tumors. Int J Radiat Biol 64:417–340
45. Peterson AV, Prentice RL, Marek P (1982) Relationship between dose of injected $^{239}Pu$ and bone sarcoma mortality in young adult beagles. Radiat Res 90(1):77–89
46. Wood D (1991) Long-term mortality and cancer risk in irradiated Rhesus monkeys. Radiat Res 126:132–140
47. Yochmowitz M, Wood D, Salmon Y (1985) Seventeen-year mortality experience of proton radiation in Macaca mulatta. Radiat Res 102:14–34
48. Maisin J, Wambersie A, Gerber G et al (1988) Life-shortening and disease incidence in C57BL mice after single and fractionated $\gamma$ and high energy neutron exposure. Radiat Res 113: 300–317
49. Ullrich RL, Storer JB (1979) Influence of gamma irradiation on the development of neoplastic disease in mice. II. Solid tumors. Radiat Res 80:317–324
50. Peterson AV, Prentice RL, Marek P (1982) Relationship between dose of injected $^{239}Pu$ and bone sarcoma mortality in young adult beagles. Radiat Res 90(1):77–89
51. Mitchel REJ, Jackson JS, McCann RA et al (1999) The adaptive response modifies latency for radiation-induced myeloid leukemia in CBA/H mice. Radiat Res 152:273–279
52. Ina Y, Tanooka H, Yamada T et al (2005) Suppression of thymic lymphoma induction by life-long low-dose-rate irradiation accompanied by immune activation in C57BL/6 mice. Radiat Res 163:153–158
53. Lacoste-Collin L, Jozan S, Cances-Lauwers V et al (2007) Effect of continuous irradiation with a very low dose of gamma rays on life span and the immune system in SJL mice prone to B-cell lymphoma. Radiat Res 168:725–732
54. Sanders CL (1996) Prevention and therapy of cancer and other common diseases: alternative and traditional approaches. Infomedix, Richland, WA, 3000pp
55. Sanders CL, Thompson RC, Bair WJ (1970) Lung cancer: dose response studies with radionuclides. Inhalation carcinogenesis, CONF-691001. NTIS, Springfield, VA, pp 285–303
56. Sanders CL, Dagle GE, Cannon WC et al (1976) Inhalation carcinogenesis of high-fired $^{239}PuO_2$ in rats. Radiat Res 68:340–360
57. Sanders CL, Lundgren D (1995) Pulmonary carcinogenesis in the F344 and Wistar rat following inhalation of $^{239}PuO_2$. Radiat Res 144:206–214
58. Sanders CL, Lauhala KE, McDonald KE, Sanders GA (1993) Lifespan studies in rats exposed to $^{239}PuO_2$ aerosol. I. Dosimetry. Health Phys 64(5):509–521

59. Sanders CL, Scott BR (2007) Smoking and hormesis as confounding factors in radiation pulmonary carcinogenesis. Dose Response 6:53–79
60. Sakai K, Hoshi Y, Nomura T et al (2003) Suppression of carcinogenic processes in mice by chronic low dose rat gamma-irradiation. Int J Low Radiat 1:142–146
61. Ootsuyama A, Tanooka H (1991) Threshold-like dose of local β irradiation repeated throughout the life span of mice for induction of skin and bone tumors. Radiat Res 125:98–101
62. Mitchel REJ, Gragtmans NJ, Morrison DP (1999) Beta-radiation-induced resistance to MNNG initiation of papilloma but not carcinoma formation in mouse skin. Radiat Res 121:180–186
63. Yamamoto O, Seyama T, Itoh H et al (1998) Oral administration of tritiated water (HTO) in mouse. III. Low dose-rate irradiation and threshold dose-rate for radiation risk. Int J Radiat Biol 73:535–541
64. Feinendegen LE (2005) Evidence for beneficial low level radiation effect and hormesis. Br J Radiol 78:3–7
65. Cohen SM, Purtilo DT, Ellwein LB (1991) Pivotal role of increased cell proliferation in human carcinogenesis. Mod Pathol 4:371–382
66. Calabrese EJ, Baldwin LA (2001) Scientific foundations of hormesis. Crit Rev Toxicol 31:351–624
67. Pollycove M, Feinendegen LE (2001) Biologic responses to low doses of ionizing radiation: detriment versus hormesis. Part 2: dose responses to organisms. J Nucl Med 42:26N–37N
68. Pollycove M, Feinendegen LE (1999) Molecular biology, epidemiology and the demise of the linear no-threshold (LNT) assumption. C R Acad Sci Paris Life Sci 322:197–204
69. Aurengo A, Averbeck D, Bonnin A et al (2005) Dose-effect relationships and estimation of the carcinogenic effects of low doses of ionizing radiation. Executive Summary. French Academy of Sciences, French National Academy of Medicine
70. Bhattacherjee G, Ito A (2001) Deceleration of carcinogenic potential by adaptation with low dose gamma irradiation. *In Vivo* 15:87–92

# 14 Animal and Human Cancer Therapeutic Studies

Readily available LDRT is needed in the U.S. for the more effective treatment of cancer

(Jerry Cuttler)

Shortly after Roentgen discovered X-rays in 1895, scientists learned that low-dose X-ray exposures cured and prevented infections and inflammations, primarily due to immune system stimulation. The stimulatory virtues of radium therapy for health and well-being were promoted until the late 1920s, when the hazards of very high radiation doses from internal radium became apparent. Studies showing that low-dose radiation is beneficial were simply ignored by the federal agencies and their advisory bodies. In 1936, a US National Academy of Sciences study discounted the known stimulatory effects of low-dose radiation [1].

Some of the many benefits of low-dose radiation identified in the early 1900s were cure of diphtheria, relief from arthritis and rheumatism pain and swelling, relief from symptoms of bronchitis, cure of gas gangrene and tuberculosis infections, and reduction of cancer incidence (in animals). *Clostridium* is the cause of gas gangrene, which is rapidly fatal if not immediately treated. During 1920s to 1940s, gas gangrene infections were successfully treated by exposure of the infected area to an X-ray dose of about 0.5 Gy. X-ray therapy often stopped the infection without requiring amputation. Mortality was cut to about 5% if patients were treated by radiotherapy prior to severe progression of the infection [2]. A book published by Kelly and Dowell in 1942 on *The Roentgen Treatment of Infections* contained hundreds of case reports demonstrating the efficacy of low doses roentgen radiation in treating various infections and cancers [3].

There has been little use of this information in medical therapies since the 1940s, except for health spas in Europe and elsewhere, where people receive therapeutic relief by inhaling naturally occurring radon gas and drinking radon water. Since the sixth century BC, the island of Icaria in the Mediterranean has been known as a health destination to relieve pain and a variety of inflammatory ailments, because of high radon levels in its hot springs [4]. People of Icaria live longer than almost anyone in the world. Radon therapy is an accepted therapy for many inflammatory diseases in Europe and an alternative therapy dismissed by most physicians in the United States [5]. Tens of thousands of people in Europe and the U.S. are exposed to radon in the air and water for treatment of inflammatory chronic diseases such as arthritis, respiratory and skin ailments, allergies, scleroderma and lupus. Radon therapy may relieve pain (most often motivation for seeking therapy) and other effects. Hundreds of

C. L. L. Sanders, *Radiation Hormesis and the Linear-No-Threshold Assumption*,
DOI: 10.1007/978-3-642-03720-7_14, © Springer Verlag Berlin Heidelberg 2010

published studies, including randomized clinical double-blind studies, have demonstrated the efficacy of radon therapy by inhalation or baths for painful inflammatory diseases [6].

EPA and BEIR VI claim that there is no safe level of radon exposure. Using the LNT assumption, the EPA has established an action level for indoor radon remediation at 4 pCi/L. However, radon levels in American radon health mines ranges from 200 to 1,600 pCi/L. Radon therapy is widely accepted as beneficial in Europe. No U.S. government agency or the American Medical Association has conducted any studies on the clinical effectiveness of radon therapy. No epidemiological study has demonstrated any cancer risk for those participating in radon therapy either in the U.S. or Europe.

Low-dose-rate exposure to low-LET radiations enhances the level of radiation adaptation, suggesting that multiple small doses may be useful in radiotherapy for cancer. Small acute doses of radiation may be used to treat cancers and prevent metastases. Combined therapy involving multiple low doses of low LET radiation with an agent(s) that sensitizes cancer cells to undergo apoptosis may be an effective strategy in treating human cancer [7]. Whole-body doses of 100–200 mGy, given to mice before implantation of tumor cells, suppressed tumor growth rate due to stimulation of NK-B cells [8]. Enhanced natural killer cells, cytotoxic T lymphocytes, IFN-$\gamma$, IL-12 and reduced GSH (Chap. 2) are associated with delayed Ehrlich solid tumor growth [9].

A study published in 1920 showed that a small X-ray dose given 7 days prior to tumor transplantation in mice decreased transplant success by over 50%, which was associated with lymphocytosis and lymphoid hyperplasia [10]. Mice were given a whole-body exposure to 75 mSv X-rays 6 h before implantation with S-180 sarcoma cells; significantly reduced tumor growth at 15 days was observed, associated with increased tumor necrosis and infiltrating lymphocytes [11]. Fractionated X-ray treatment of human colorectal tumor clones elicited an inducible-like radioprotective response [12].

Ionizing radiation can reduce tumor growth outside the field of radiation; this is known as the abscopal effect. Mice bearing a syngeneic mammary carcinoma in both flanks were treated with the growth factor Flt3-L for 10 days after local radiation therapy with a single dose of 2 or 6 Gy to one flank. Tumor growth of the shielded flank tumor was depressed. No growth depression was observed in T-cell-deficient (nude) mice. This indicates that the abscopal effect is in part mediated by T-cell immunity [13].

Low-dose whole-body irradiation, substituted for focused high-dose fractions, significantly improved survival in mice in the Lewis lung cancer model over conventional radiotherapy (Fig. 14.1) [15]. TBI (total body irradiation) doses of 0.1–0.4 Gy significantly suppressed lung metastases in mice (Figs. 14.2 and 14.3) [16]. Metastases in lung and lymph nodes were reduced by up to 70% in rats receiving 200 mGy X-rays TBI at day 14 after implantation of hepatoma cells [17]. Doses of 10, 50, 100 and 2,000 mGy $\gamma$-rays given 24 h before subcutaneous glioma 261 tumor cells significantly inhibited tumor growth [18].

The number of B16 melanoma metastatic foci in mouse lung was significantly reduced by administration of radon in drinking water provided the number of tumor cells was small and the radon treatment was started before tumor cell administration [19]. A whole-body dose of 0.5 Gy $\gamma$-rays significantly suppressed pulmonary colony formation from injected B16 cells, associated with suppression of matrix metalloproteinase-2, a promoter of metastases [20, 21]. Induction of thymic lymphoma in C57BL/6 mice by acute X-irradiation was suppressed by continuous low-dose $\gamma$-irradiation [22].

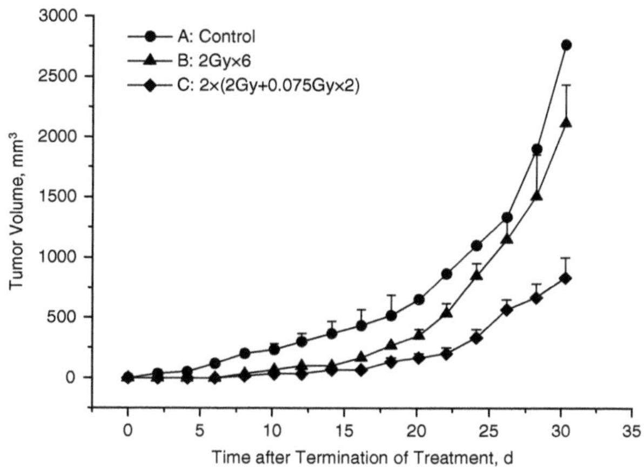


**Fig. 14.1** Growth curves of Lewis lung cancer following 2 Gy fractions (0.3 Gy/min) and 75 mGy TBI (12.5 mGy/min) [14]

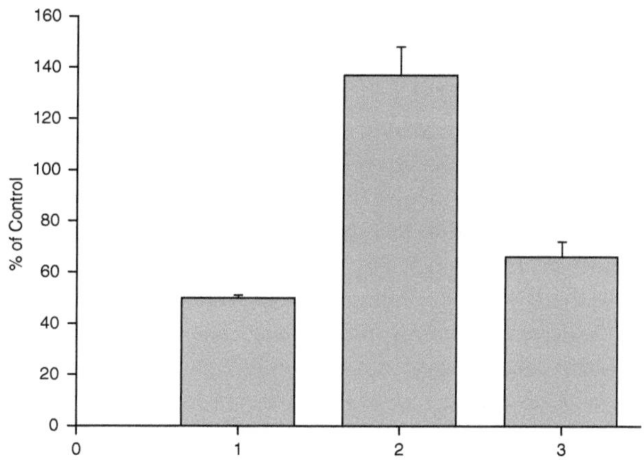


**Fig. 14.2** Effect of 75 mGy TBI on tumor size, survival time and mortality rate in C57BL/6J mice implanted with Lewis lung cancer cells compared with unirradiated controls. (1) Tumor size; (2) survival time; (3) mortality rate 19

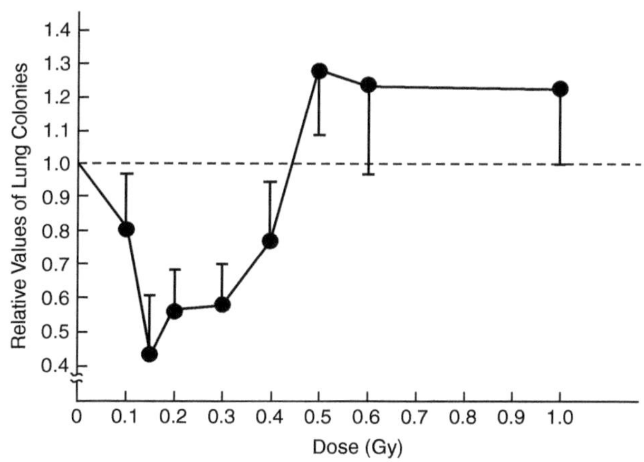


**Fig. 14.3** Effect of TBI in mice given 12 days after transplantation of Lewis lung cancer cells. Lung colonies were counted at 20 days after TBI. Values are relative to that obtained for unirradiated mice 1

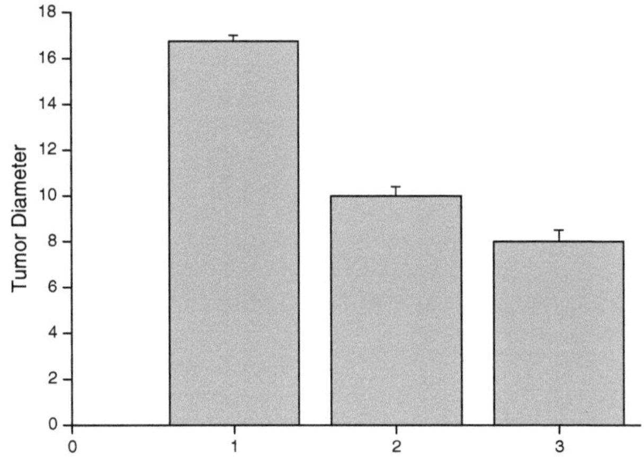


**Fig. 14.4** Effect of low-dose radiation and chemotherapy with mitomycin C on Lewis lung cancer growth. (1) Sixteen days after tumor implantation; (2) 3 mg/kg i.p. MMC given on day 10; (3) TBI with 75 mGy given at 6 h before 3 mg/kg i.p. MMC 19

Low-dose irradiation enhances the therapeutic response of cancers to chemotherapeutic agents. For example, low-dose TBI irradiation (75 mGy) given 6 h before mitomycin C significantly reduced Lewis lung cancer growth below that observed with mitomycin C alone (Fig. 14.4) [23].

There is a considerable amount of interest on the application of apoptosis in oncology, both in the pharmaceutical development of new apoptotic drugs and in their evaluation in clinical trials [24]. However, there is currently a great reluctance to investigate the use of low-dose irradiation therapies that have shown evidence of improving the outcome of cancer treatments [25, 26]. Physicians have concerns about the ethics of providing low-dose therapy for preventing or curing cancer.

A major concern regarding breast cancer, prostate cancer and colon cancer is the early detection and removal of tumors before metastases can break off and lodge in a vital organ or in nearby bone tissue. Prostate cancer is the second most common cancer in North American men and second leading cause of cancer mortality for those older than 65 years. Early detection has not been shown to decrease prostate cancer mortality for those over the age of 50 years. Cuttler suggested that older men should be given periodic low-dose, total-body irradiation therapy to prevent and control cancers, such as prostate cancer [27]. Such doses exhibit no side-effects and could be delivered by a whole-body CT scan.

Repeated low-dose, low-toxicity chemotherapy administered over a prolonged period of time, called Low-dose metronomic therapy (LDMT) using anti-angiogenic drugs, has been shown to be effective in breast and ovarian cancers and in advanced cancers [28]. LDMT has similarities to hormesis. Likewise, hormetic effects are induced by various foods [29].

It is well known that low doses of radiation stimulate the immune system, which can destroy metastases [30]. Total or half-body low-dose irradiation should be an essential part of any cancer treatment [1]. Several trials have been carried out over the past 30 years that have all shown positive results. Positive palliative results were found in advanced cases of leukemia treated in the 1950s with whole-body or hemi-body exposure to fractionated radiation up to a total dose of 2 Gy [32]. Patients tolerate the low-dose therapy very well. Low-dose, low-dose-rate radioimmunotherapy, using β-irradiation, has been used to successfully treat follicular lymphoma. Ovarian cancer and colon cancer metastases were successfully eliminated from two cancer patients at Tohoku University by LDRT (Low-Dose Radiation Therapy) [1].

**Table 14.1** Low-dose radiation schedule for treatment of non-Hodgkin's lymphoma [16, 34]

| | Low-dose radiation therapy | |
|---|---|---|
| Radiation field | Whole body | Half body |
| Total dose | 1.5 Gy | |
| Fractionated dose schedule | 0.10 Gy | 0.15 Gy |
| | 3 Times a week | 2 Times a week |
| | 15 Times | 10 Times |

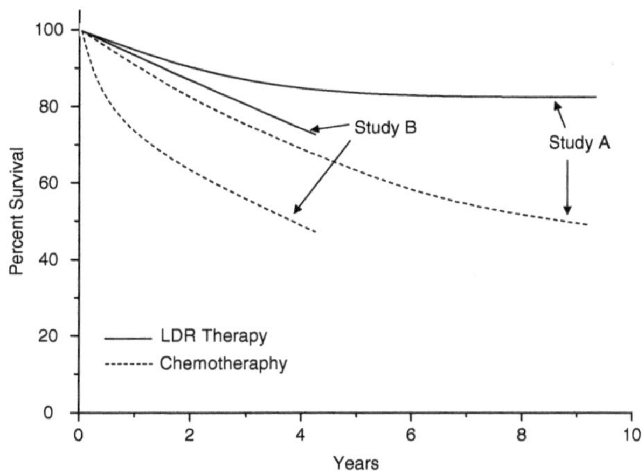


**Fig. 14.5** Comparison of traditional four-drug chemotherapy of LDRT with TBI or HBI in the treatment of non-Hodgkin's lymphoma (Study A [34] and Study B [16])

Clinical trials with LDRT are being carried out in China by Dr. Shu-Zheng Liu [31]. Individual patients with various types of cancer are receiving LDRT from a few radiation oncologists in Canada and the U.S. that are willing to do it. Good responses in these patients will hopefully encourage other physicians to adopt LDRT so that it will be more available to cancer patients. Use of low-dose irradiation, in conjunction with chemotherapy, provides a better cure rate than chemotherapy alone for non-Hodgkin's lymphoma. These results were obtained from clinical trials at Harvard University in the US in 1976 and 1979, at Tohoku University in Japan in 1997 and at Bergonie Institute in France in 1998 [33–36]. Multiple myeloma has been successfully treated with low-dose TBI [37]. A whole-body dose of <250 mGy is effective in the treatment of lymphosarcoma [33]. A total dose of 2 Gy given in 100–200 mGy fractions (5 fractions/week) is effective in the treatment of relapsed/refractory Hodgkin's lymphoma [38].

Administration of low-dose TBI therapy or upper half-body irradiation (HBI) therapy (6 MV X-rays) to patients with non-Hodgkin's lymphoma, who had previously received chemotherapy and/or high-dose radiation of tumors has been reported [16, 33–35]. TBI/HBI therapy consists of individual fractions of 10–15 cGy at 30 cGy/week for 5 weeks to give a total dose of 150 cGy (Table 14.1). TBI/HBI therapy is well tolerated with no symptomatic side effects. TBI therapy was reviewed in 2000 [36].

The Harvard studies reported a 4-year survival of 70 and 74% with added LDRT as compared to 40 and 52% for those treated only with multi-drug chemotherapy [33, 34]. HBI was as effective as TBI at Tohoku University, where 4-year survival was 84% after added LDRT compared to 65% for those who received multi-drug chemotherapy and/or high-dose regional radiotherapy (Fig. 14.5). The 12-year survival of the TBI/HBI patients continued at 84% [16]. The response rate at the Bergonie Institut to low-dose TBI was 83%, extending recurrence-free survival [35].

## References


1. Cuttler JM (2008) Therapeutic applications of radiation hormesis. Cuttler & Associates, Mississauga, ON
2. Cuttler JM (2004) Low-dose irradiation therapy to cure gas gangrene infections. Int J Low Radiat 1:318–28
3. Cuttler JM (2008) Book review: Roentgen treatment of infections by JF Kelly and DA Dowell. Can Nucl Soc Bull 29:43–4
4. http://www.cnn.com/2009/HEALTH/04/16/longevity/index.html
5. Erickson BE (2007) The therapeutic use of radon: a biomedical treatment in Europe; an "alternative" remedy in the United States. Dose Response 5:48–62
6. Becker K (2004) One century of radon therapy. Int J Low Radiat 1:333–57
7. Scott BR, Belinsky SA, Leng S et al (2009) Radiation-simulated epigenetic reprogramming of adaptive-response genes in the lung-an evolutionary gift for mounting adaptive protection against lung cancer. Dose Response 7:104–131
8. Cheda A, Wrembel-Wargocka J, Lisiak E et al (2004) Single low doses of X rays inhibit the development of experimental tumor metastases and trigger the activities of NK cells in mice. Radiat Res 161:335–40
9. Hayase H, Ohshima Y, Mareyuki M, Kojima S (2008) The enhancement of Th-1 immunity and the suppression of tumour growth by low dose g-radiation. Int J Low Radiat 5:275–89
10. Murphy JB (1920) The effect of physical agents on the resistance of mice to cancer. Proc Natl Acad Sci U S A 6:35–8
11. Yu H-S, Song A-Q, Lu Y-D et al (2004) Effects of low-dose radiation on tumor growth, erythrocyte immune function and SOD activity in tumor-bearing mice. Chin Med J 117:1036–9
12. Qutob SS, Asha AS, Pathak S et al (2006) Fractionated X-radiation treatment can elicit an inducible-like radioprotective response that is not dependent on the intrinsic cellular X-radiation resistance/sensitivity. Radiat Res 166:590–9
13. Demaria S, Bruce NG, Devitt ML et al (2004) Ionizing radiation inhibition of distant untreated tumors (abscopal effect) is immune mediated. Int J Radiat Oncol Biol Phys 58:862–70
14. Jin GH, Jin SZ, Liu Y et al (2005) Therapeutic effect of gene therapy in combination with local x-irradiation in a mouse malignant melanoma model. Biochem Biophys Res Commun 330:975–81
15. Wu N, Jin S-Z, Pan X-N, Liu S-Z (2008) Increase in efficacy of cancer radiotherapy by combination with whole-body low dose irradiation. Int J Radiat Biol 84:201–10
16. Sakamoto K, Myogin M, Hosoi Y et al (1997) Fundamental and clinical studies on cancer control with total or upper half body irradiation. J Jpn Soc Ther Radiol Oncol 9:161–75
17. Hashimoto S, Shirato H, Hosokawa M et al (1999) The suppression of metastases and the change in host immune response after low-dose total-body irradiation in tumor-bearing rats. Radiat Res 151:717–24
18. Lumniczky K, Szatmari T, Bogdandi N, Safrany G (2007) Effects of low dose irradiation on the main immune parameters and on the antitumor immune surveillance in mice. In: Sixth LOWRAD Conference, Budapest, Hungary, Abstract; p 78
19. Takahashi M, Kojima S (2006) Suppression of atopic dermatitis and tumor metastasis in mice by small amounts of radon. Radiat Res 165:337–42
20. Ohshima Y, Tukimoto M, Kojima S (2007) Inhibitory mechanism of low-dose, whole-body

irradiation with gamma-rays against tumor metastasis. In: Sixth LOWRAD Conference, Budapest, Hungary, Abstract; p 103
21. Ohsma Y, Tsukimoto M, Kojima S (2008) The novel mechanism of metastasis inhibition by low-dose whole-body irradiation with gamma-rays. Int J Low Radiat 5:156–67
22. Ina Y, Tanooka H, Yamada T et al (2005) Suppression of thymic lymphoma induction by life-long low-dose-rate irradiation accompanied by immune activation in C57BL/6 mice. Radiat Res 163:153–8
23. Liu S-Z (2007) Cancer control related to stimulation of immunity by low-dose radiation. Dose Response 5:39–47
24. http://www.researchandmarkets.com/products/a3992c/analytical_tool_apoptosis_in_oncology
25. Safwat A (2000) The role of low-dose total body irradiation in treatment of non-Hodgkin's lymphoma: a new look at an old method. Radiother Oncol 56:1–8
26. Cuttler JM, Pollycove M, Welsh JS (2000) Application of low doses of radiation for curing cancer. Can Nucl Soc Bull 21:45–6
27. Cuttler JM (2006) Low-dose irradiation for controlling prostate cancer. Int J Low Radiat 2:45–59
28. Satti J (2009) The emerging low-dose therapy for advanced cancers. Dose Response (in press)
29. Mattson MP, Cheng A (2006) Neurohormetic phytochemicals: low-dose toxins that induce adaptive neuronal stress responses. Trends Neurosci 29:632–9
30. Wojcik M, Zabek M, Rzeznik D et al (2002) Half body irradiation (HBI) in palliative treatment of multiple cancer metastases – contemporary evaluation. Wspolczesna Onkologia 8:395–9
31. Personal communication from Jerry M. Cuttler MD, Cuttler & Associates Inc., 1781 Medallion Court, Mississauga, Ontario, L5J 2L6, Canada. Contact information is: Liu Shu-Zheng MD, Professor Emeritus, Radiation Biology, Jilin University Health Sciences Center, 828 Xinmin street, Changchun 130021, P.R. China, Email: drliusz@yahoo.com
32. Jaworowski Z (2009) Radiation hormesis – a remedy for fear. BELLE Newsletter 15:14–20
33. Chaffey JT, Rosenthal DS, Moloney WD, Hellman S (1976) Total body irradiation as treatment for lymphosarcoma. Int J Radiat Oncol Biol Phys 1:399–405
34. Choi NC, Timothy AR, Kaufman SD et al (1979) Low dose fractionated whole body irradiation in the treatment of advanced non-Hodgkin's lymphoma. Cancer 43:1636–42
35. Richaud PM, Soubeyran P, Eghbali H et al (1998) Place of low-dose total body irradiation in the treatment of localized follicular non-Hodgkin's lymphoma: results of a pilot study. Int J Radiat Oncol Biol Phys 40:387–90
36. Safwat A (2000) The role of low-dose total body irradiation in treatment of non-Hodgkin's lymphoma: a new look at an old method. Radiother Oncol 56:1–8
37. Holder DL (1965) Total body irradiation in multiple myeloma. Radiology 84:83–6
38. Mishra KP, Ahmed M, Hill RP (2008) Low-dose radiation effects on human health with implications to radioprotection and cancer radiotherapy. Int J Radiat Biol 84:441–4

# Conclusions, Summary, and Importance 15

The hormetic model is not an exception to the rule – it is the rule

*(Edward Calabrese)*

The precautionary principle [1] is invoked by LNT supporters in the belief that low-dose radiation causes irreversible harm to the public. LNT protagonists conclude that there is an absence of a scientific consensus and places the burden of proof on advocates of radiation hormesis. They then proceed to ignore the overwhelming body of proof that the LNT assumption is diametrically wrong in predicting risk of cancer at low doses, and continue to unrealistically lower radiation standards with a conscience unwilling to examine the myriad benefits of low dose, low dose-rate, low LET ionizing radiation.

The LNT model is scientifically indefensible, yet is still used because of its simplicity and convenience [2,103]. "In order to make them believe the LNT dogma, radiobiologists have consistently misled students, physicians, professors, the media, the public, government advisory boards, and heads of nations" [3]. Radiation protection is in the pay of the LNT and the price that we are paying in terms of human life, resources and money is very high. The unethical behavior and pronouncements of many radioprotection groups and individuals needs to be questioned and openly debated [4, 5].

Policies and myths promulgated nearly 50 years ago after WWII by the ICRP and later by other agencies have convinced most people that any amount of ionizing radiation is dangerous. "Unfortunately, the initially quite reasonable ICRP recommendations have degenerated over the decades into applications of the LNT/Collective Dose dogma" [6]. Calabrese, in a fascinating discussion of the historical origins of the LNT assumption, challenges the validity of extrapolation from high dose to low dose by early evidence in the 1920s. Over time the linearity concept was generalized from genetic effects to somatic effects induced by radiation and carcinogenic chemicals [7]. Experimental and epidemiological studies have challenged the validity of the LNT assumption, strongly indicating the presence of a threshold and/or benefits (hormesis) from low doses of low-LET ionizing radiation. The assessment of radiation carcinogenesis at low doses, using the LNT assumption, is flawed because it assumes that the mechanisms which lead to cancer formation are identical at low and high doses. False assumptions and projections of cancer risk derived from linear extrapolation of harmful effects at high doses are the result of using the LNT assumption [8–10].

C. L. L. Sanders, *Radiation Hormesis and the Linear-No-Threshold Assumption*,
DOI: 10.1007/978-3-642-03720-7_15, 

Fear of radiation is common in medicine and in the nuclear industry. The result of radiophobia is a frightened public due to exaggerated adverse health claims by professionals in radiation protection [4]. Physicians and patients avoid exposures from diagnostic radiological examinations out of fear of the carcinogenic effects of ionizing radiation. It is widely recognized that patients in radiology and nuclear medicine are fearful about future cancer risk. Recent LNT-related estimates of cancer risk from CT scans using the LNT assumption have enhanced this fear [11, 12]. Actually, CT scans are likely to decrease cancer risk [13].

There was a heavy price to pay due to the response from Chernobyl that had nothing directly due to do with radiation risk but from unwarranted fear of radiation risk. The human Chernobyl dose was a 100 times less than the background dose seen in parts of the world, a dose shown not to cause any adverse health effects but beneficial effects. The costs from Chernobyl to Western Europe, Ukraine and Belarus are estimated at ~US $ half a trillion, most of which was wasted on efforts to control insignificant radiation exposures [14]. There were an estimated 1,250 excess suicides and >100,000 excess elective abortions, while 200,000 individuals experienced an unnecessary traumatic evacuation. There was a decrease, not an increase, in birth defects and cancer in those exposed to fallout. The Chernobyl experience should be deemed *a deeply immoral use of our scientific heritage*, a comment made by the late Lauriston Taylor, the first president of the NCRP, when referring to LNT-promulgated radiation standards.

Radiation-dose thresholds for cancer formation are common in both animal and human studies [15–35]. Radiation prevention of cancer from exposure to $\gamma$- or X-irradiation is observed at dose rates from 0.5 to 10 mGy/day. The beneficial effects are partially or wholly balanced by the harmful effects at dose rates of 10–100 mGy/day [36, 37]. Cancer incidence thresholds of 0.1–3.0 Gy, following acute, whole-body exposures to low-LET radiations, have been observed from dose–response data of animal and human studies [17, 31]. There is no evidence of radiation-induced cancer in humans at doses <100 mSv from acute doses or at doses <500 mSv for protracted doses [38]. Thresholds for cancer were common in A-bomb survivors [15, 39–43]. These include thresholds of 200 mSv for solid cancer mortality in both cities [15]. The threshold for leukemia in all A-bomb survivors was < 300 mSv [39–41]. Meta-analyses have shown a threshold of 1–2 Gy of low-LET radiation for lung cancer in never smokers or smokers [18, 44, 45]; the data indicated "a reduction of the natural incidence" [45].

Alpha-dose thresholds of 1–10 Gy have been found for bone tumors from radium [19], liver tumors from Thorotrast and plutonium [46, 47], and lung cancer from plutonium [48]. This amounts to thresholds of 10–100 Sv using an RBE of 10. Proponents of the LNT believe that tumors are monoclonal and develop from the offspring of a single genetically damaged cell by a single radiation hit, which is then followed by a series of genetic and cytogenetic alterations leading to cancer [49]. In cases of high-dose thresholds, about half of all target cells are irradiated by $\alpha$-particles before any bone or liver tumors are found, disproving the single cell, single hit model. The threshold for lung cancer in uranium miners is as high as 800 WLM or about 4 Sv [20, 50].

The beneficial effects of low doses of ionizing radiation have been reported in over 3,000 publications during the last century [51, 52]. The hormesis model is based on extensive evidence following exposure of plants, fungi, insects, mammals and humans to a variety of chemicals and ionizing radiation [3, 51–60]. Recently, an entire issue of the American Journal of Pharmacology and Toxicology was given to hormesis [105]. The biphasic dose–response model uses low-dose modeling and recognizes both beneficial (stimulation at low-dose) and harmful (inhibition at high-dose) responses for physical or chemical agents. The model was named "hormesis" after the Greek word "to excite" following observations of scientists at the University of Idaho in 1943, who observed a U-shaped dose–response curve for fungi exposed to red cedar extracts [61].

Life is not possible without DNA repair of oxygen-related damage and the elimination of DNA-damaged cells before they can cause life-threatening diseases. The rate of DNA damage caused by background and low doses of ionizing radiation is exceedingly small compared to DNA damage caused by breathing oxygen (~500 g $O_2$ per day for standard man). The radio-adaptive response is associated with stimulation of protective mechanisms (antioxidant defenses, DNA repair, immune defenses and apoptosis), leading to a reduction in naturally (mostly $O_2$ related) occurring damage, including cancer [17, 62].

Cellular stimulatory effects are observed following radiation doses of 1–500 mSv, and damaging effects at doses >500 mSv. Low-dose, low-dose-rate, low-LET ionizing radiation: protects against chromosome aberration formation from a following high dose [63–65]; protects against mutations from a high-radiation dose given either before or after a high dose [66, 67]; decreases precancerous (transformed) cells [68, 69]; enhances immune function [70, 71]; suppresses induced and spontaneous cancers [16, 72]; decreases cancer metastases [73, 74] and decreases prevalence of many noncancer diseases [75]. Apoptosis of transformed cells may be activated by low-dose, low-LET gamma or X-radiations, removing cigarette-smoke and high-LET radiation induced, genetically damaged cells before they can develop into cancers [18].

Important well-regarded research published in prestigious scientific journals often later turns out to be false for a variety of reasons [76]. Proponents of the LNT assumption routinely exclude data from a vast array of peer-reviewed scientific literature and then statistically challenge the robustness of low-dose data, while in the process ignoring data from large populations (such as inhabitants of high background dose) that demonstrate hormesis [62, 77]. They claim that they are being prudent by avoiding low doses (since they believe all doses are harmful). They use the collective-dose fallacy, even though they know that it is scientifically invalid. They claim that the LNT assumption is simple to use and that threshold models along with hormesis are too complicated, that dropping the LNT would complicate the regulation of radiation.

Biased reporting of single and pooled epidemiological studies is a significant problem in clarifying causational relationships. Biased data selection, inappropriate manipulation of data, lumping of all low-dose data into one dose bin, dose lagging (ignoring some of the radiation dose), including individuals who receive low-dose radiation in unexposed groups, not adjusting for important confounding factors (particularly cigarette smoking), using linear extrapolation from high to low doses over wide-dose intervals and ignoring radiation hormesis for no other reason than it does not fit the LNT assumption [78].

A negative cancer risk has been found in populations living in regions of high background radiation, in nuclear and medical workers and following accidental exposures [62, 79]. A highly, statistically significant, negative association between radon concentration and lung cancer was found in the United States, even after adjusting for smoking and over 50 other categorizations [80]. Sequential determinations of cancer mortality have consistently shown evidence of decreased cancer in UKAEA nuclear workers from 1946 to 1997 [81] and in Chernobyl liquidators from 1990 to 1998 [82]. Three Scandinavian studies showed that $^{131}I$ used for both diagnosis and therapy of hyperthyroidism decreased thyroid cancer and other cancer incidences [83–85].

Ionizing radiation is essential for life and is needed for optimal health. Many people live in radiation-deficient environments. In fact, the zeal of radiation protection organizations can, and probably already has, led to radiation deficiency. Cohen showed that radon deficiency enhances the risk of lung cancer [80]. A recent well-performed, case–control study of indoor radon showed that exposure to radon at the EPA action level is associated with a decrease of ~60% in lung cancer [86]. Lung cancer deaths were unusual in Saxony, Germany in the nineteenth century before cigarette smoking, even though radon levels reached 100 times the current USEPA remediation standard [53]. The optimum radiation dose of chronic radiation may be 60 mSv/year [87]. This compares to an average world background dose of 2.5 mSv/year. Luckey predicted that 15 mSv/year would eliminate most cancer mortality [53].

The Chernobyl reactor accident on 26 April 1986, was the worst nuclear power accident in history. No health problems, other than the acute radiation syndrome in about 100 initial firefighters, were found in populations living in contaminated regions around the Chernobyl reactor that can be directly attributed to radiation exposure. There was no excess mortality from thyroid cancer or leukemia, no excess cataracts, congenital abnormalities or infant mortality. In fact, the quarter million Chernobyl liquidators, who received a mean dose of about 100 mSv experienced a 15–30% reduction in cancer deaths [88].

Cancer risk estimates using the LNT assumption are largely based on studies of Japanese A-bomb survivors who experienced excess cancer only at doses >100 mSv [12]. A total of 86,543 persons were exposed in Hiroshima and Nagasaki, of which 45,148 received doses <10 mSv. Over 90% of the exposed population received doses <500 mSv. To those receiving a single instantaneous dose of <1 Sv, there was a lifetime of improved health [3, 43]. No statistically significant effects have been found in birth defects, neonatal deaths, stillbirths, leukemia, solid cancers, death rates in offspring, sex ratio, growth and development during childhood, mental retardation, chromosomal aneuploidy and translocations or mutations in those receiving doses of <100 mSv [3, 89, 90]. Japanese A-bomb survivors are living longer than unexposed controls, who received the same long-term medical care and other benefits as the exposed cohorts. Noncancer mortality rates for Nagasaki survivors was 65% of age-matched controls ($p < 0.05$) [43]. All cause mortality rates in Nagasaki and Hiroshima combined showed a significant decreased relative risk ($p < 0.01$) for those exposed to 140 mSv [91]. Correction for "special medical care of survivors" failed to remove the beneficial response [43].

The "healthy worker effect" (HWE) has been used in an attempt to explain less than the expected cancer rates in epidemiological studies of irradiated populations [92]. Many researchers consider the HWE to be useless in epidemiological studies [93, 94]. Annual medical physicals have no significant impact on cancer mortality [21, 81, 95, 96]. A recent study of about 600,000 nuclear workers in 15 countries found a large reduction in all cause and all cancer mortality (mean PROFAC of 38 and 26%, respectively). The authors attributed this large effect to the HWE and totally disregarded the possibility of a benefit [35, 97, 98]. The ubiquitous nature of the radio-adaptive response in cellular and animal studies, the large hormesis effect in nuclear and radiological workers and the presence of hormesis in non-worker populations and in radiation workers when compared to unirradiated workers in the same facilities negates the HWE as an explanation for radiation hormesis [18, 52, 62, 63].

About 10,000 people occupied 1,700 apartments in Taiwan for up to 20 years, receiving a cumulative average radiation dose of 0.4 Sv from construction steel contaminated with $^{60}$Co [16]. A sharply reduced incidence of cancer and birth defects was seen in apartment inhabitants. They coincidentally accomplished the almost perfect study in a human population that demonstrates the highly significant protective effects of near-continuous exposure to gamma radiation [99]. In fact, the very low cancer mortality rate may have been due to the therapeutic effects of LDRT-like exposure on spontaneously developing cancers.

Nuclear power is safe, inexpensive, reliable, ecologically beneficial, healthy, and climatically neutral. Japan has a high percentage of power generation from nuclear power plants. Many Japanese physicians accept the fact that low-dose ionizing radiation is beneficial and have radiation therapy clinics that prescribe low doses for various diseases [53]. The presence of hormesis has far-reaching implications for cancer therapy [100]. LDRT (Low-dose radiation therapy) with 100 mGy × 15 exposures for a total dose of 1.5 Gy given over 5 weeks has shown in two clinical trials to be very effective in treating non-Hodgkin's lymphoma [73, 101]. Cumulating anecdotal accounts speak of the efficacy of LDRT for a variety of other cancer types. Jerry Cuttler has been a proponent of LDRT for cancer during the last decade [104]. Dr. Shu-Zheng Liu in China has been carrying out a clinical trial for a variety of cancers using LDRT. That such successful therapy for cancer has not been used alone or in combination with chemotherapy and high-dose radiotherapy is largely due to ignorance of radiation oncologists and/or prejudice from following the LNT assumption.

Political and vested interests are often behind the exclusion of radiation hormesis in setting radiation protection standards. Failure to study or even acknowledge the presence of radiation hormesis is a result of political influences within the radiation protection community [102]. Adoption of current radioprotection standards based upon application of the LNT assumption does not protect against diseases at low doses but results in an increased risk. Proponents of the LNT say that there is nothing new about the studies on hormesis and that we must study more. They consistently reject and deliberately ignore unwanted data. They claim that radiation hormesis has not been seriously challenged in scientific peer-reviewed literature. They falsely claim that there is no reliable data at doses <100 mSv, and then ignore the abundant data that is available. They apparently hope to continue this mantra before their careers are over and their deception has been fully revealed. It is time to use common sense.

## References

1. http://en.wikipedia.org/wiki/Precautionary_principle
2. Scott BR (2008) It's time for a new low-dose-radiation risk assessment paradigm – one that acknowledges hormesis. Dose Response 6:333–51
3. Luckey TD (2008) Atomic bomb health benefits. Dose Response 6:369–82

4. Cuttler JM (2007) What becomes of nuclear risk assessment in light of radiation hormesis? Dose Response 5:80–90
5. Jaworowski Z (1999) Radiation risk and ethics. Phys Today 52:24–29
6. Becker K (2006) Regulatory low dose limits: from science to political correctness? Int J Low Radiat 3:159–65
7. Calabrese EJ (2009) The road to linearity: why linearity at low doses became the basis for carcinogen risk assessment. Arch Toxicol 83:203–25
8. Tubiana M, Aurengo A, Averbeck D et al (2005) Dose-effect relationships and the estimation of the carcinogenic effects of low doses of ionizing radiation. Joint Report of the French National Academies of Science and of Medicine, Edition Nucleon, Paris
9. Cuttler JM, Pollycove M (2009) Nuclear energy and health. Dose Response 7:52–89
10. Calabrese EJ (2009) Getting the dose-response wrong: why hormesis became marginalized and the threshold model accepted. Arch Toxicol 83:227–47
11. Griffey RT, Sodickson A (2009) Cumulative radiation exposure and cancer risk estimates in emergency department patients undergoing repeat or multiple CT. AJR 192:887–92
12. Brenner DJ, Hall EJ (2007) Computed tomography – an increasing source of radiation exposure. N Engl J Med 357:2277–84
13. Scott BR, Sanders CL, Mitchel REJ, Boreham DR (2008) CT scans may reduce rather than increase the risk of cancer. J Am Phys Surg 13:7–10
14. Jaworowski A (2008) The paradigm that failed. Int J Low Radiat 5:151–5
15. Shimitzu Y, Kato H, Schull WJ (1990) Studies on the mortality of A-bomb survivors. 9. Mortality, 1950–1985; Part 2. Cancer mortality based on the recently revised doses (DS86). Radiat Res 121:120–41
16. Chen WL, Luan YC, Shieh MC et al (2007) Effects of cobalt-60 exposure on health of Taiwan residents suggest new approach needed in radiation protection. Dose Response 5:63–75
17. Duport P (2003) A database of cancer induction by low dose radiation in mammals: overview and initial observations. Int J Low Radiat 1:120–31
18. Sanders CL, Scott BR (2008) Smoking and hormesis as confounding factors in radiation pulmonary carcinogenesis. Dose Response 6:53–79
19. Rowlands RE, Stheney AF, Lucas HF (1983) Dose-response relationships for radium-induced bone sarcomas. Health Phys 44(Suppl 1):15–31
20. Bruske-Hohfeld I, Rosario AS, Wolke G et al (2006) Lung cancer risk among former uranium miners of the Wismut Company in Germany. Health Phys 90:208–16
21. Berrington A, Darby SC, Weiss HA et al (2001) 100 years of observation on British radiologists: mortality from cancer and other causes 1987–1997. Br J Radiol 74:507–19
22 Tubiana M (2009) Can we reduce the incidence of second primary malignancies occurring after radiotherapy? A critical review. Radiother Oncol 91:1
23. Preston DL, Ron E, Tokuoka S et al (2007) Solid cancer incidence in atomic bomb survivors: 1958–1998. Radiat Res 168:1–64
24. Little MP, Muirhead CR (2000) Derivation of low dose extrapolation factors from analysis of the curvature in the cancer incidence dose response in Japanese atomic bomb survivors. Int J Radiat Biol 76:939–53
25. Mortazavi SMJ, Ghiassi-Nejad M, Rezaiean M (2005) Cancer risk due to exposure to high levels of natural radon in the inhabitants of Ramsar, Iran. Int Cong Ser 1276:436–7
26. Mays CW, Spiess H, Gerspach A (1978) Skeletal effects following 224Ra injections into humans. Health Phys 35:83–90
27. Gilliland FD, Hunt WC, Archer VE et al (2000) Radon progeny exposure and lung cancer risk among non-smoking uranium miners. Health Phys 79:365–72
28. Roscoe RJ, Steenland K, Halperin WE (1989) Lung cancer mortality among nonsmoking uranium miners exposed to radon daughters. J Am Med Assoc 262:629–33

29. Kostyuchenko VA, Krestina LYu (1994) Long-term irradiation effects in the population evacuated from the East-Urals radioactive trace area. Sci Total Environ 142:119–25
30. Ivanov VK, Gorski AI, Maksioutov MA et al (2001) Mortality among the Chernobyl emergency workers: estimation of radiation risks (preliminary analysis). Health Phys 81:514–21
31. Tanooka H (2004) Threshold dose problems in radiation carcinogenesis: a review of non-tumour doses. Int J Low Radiat 1:329–33
32. Davis FG, Boice JD, Hrubec Z et al (1989) Cancer mortality in a radiation-exposed cohort of Massachusetts tuberculosis patients. Cancer Res 49:6130–6
33. Howe GR (1995) Lung cancer mortality between 1950 and 1987 after exposure to fractionated moderate-dose-rate ionizing radiation in the Canadian fluoroscopy chart study and a comparison with lung cancer mortality in the atomic bomb survivor study. Radiat Res 142:295–305
34. Franklyn JA, Maisonneuve P, Sheppard M et al (1999) Cancer incidence and mortality after radioiodine treatment for hyperthyroidism: a population-based cohort study. Lancet 353:2111–5
35. Vrijheid M, Cardis E, Blettner M et al (2007) The 15-country collaborative study of cancer risk among radiation workers in the nuclear industry: design, epidemiological methods and descriptive results. Radiat Res 167:361–79
36. Gregoire O, Cleland MR (2006) Novel approach to analyzing the carcinogenic effect of ionizing radiation. Int J Radiat Biol 82:13–9
37. Okamoto K (1987) Critical values of linear energy transfer, dose rates and doses for radiation hormesis. Health Phys 52:671–4
38. Tubiana M, Feinendegen LE, Yang C, Kaminski JM (2009) The linear no-threshold relationship is inconsistent with radiation biologic and experimental data. Radiology 251:13–22
39. BEIR I Committee. (1972) The effect on populations of exposure to low level of ionizing radiation. National Academy Press, Washington, DC
40. Land CE (1980) Estimating cancer risks from low doses of ionizing radiation. Science 209:1197–203
41. Shimizu Y, Kato H, Schull WJ, Mabuchi K (1992) Dose-response analysis among atomic-bomb survivors exposed to low-level radiation. In: Gugahara T, Sagan L, Y Aoyama (eds) Low dose irradiation and biologic defense mechanisms. Excerpta Medica, Amsterdam, pp 71–4
42. Mettler FA, Moseley RD (1985) Medical effects of ionizing radiation. Grune and Stratton, Orlando
43. Kondo S (1993) Health effects of low level radiation. Kinki University Press, Osaka, Japan
44. Enflo A (2002) Lung cancer risks from residential radon among smokers and non-smokers. J Radiol Prot 22:A95–9
45. Rossi HH, Zaider M (1997) Radiogenic lung cancer: the effects of low doses of low linear energy transfer (LET) radiation. Radiat Environ Biophys 36:85–8
46. Tokarskaya ZB, Zhuntova GV, Scott BR et al (2006) Influence of alpha and gamma radiations and non-radiation risk factors on the incidence of malignant liver tumors among Mayak workers. Health Phys 91:296–310
47. Van Kaick G, Wesc H, Luhrs H et al (1991) Neoplastic diseases induced by chronic alpha irradiation. Epidemiological, biophysical and clinical results by the German Thoratrast study. J Radiat Res 32:20–33
48. Tokarskaya ZB, Okladnikova ND, Belyaeva ZD et al (1997) Multifactorial analysis of lung cancer dose-response relationships for workers at the Mayak nuclear enterprise. Health Phys 73:899–905
49. Michor F, Iwasa Y, Nowak M (2004) Dynamics of cancer progression. Nat Rev Cancer 4:197–205
50. Xiang-Zhen X, Lubin JH, Jun-Yao L et al (1993) A cohort study in southern China of tin miners exposed to radon and radon decay products. Health Phys 64:120–31
51. Luckey TD (1980) Hormesis with ionizing radiation. CRC, Boca Raton, FL

52. Luckey TD (1991) Radiation hormesis. CRC, Boca Raton, FL
53. Luckey TD (2007) Documented optimum and threshold for ionizing radiation. Int J Nucl Law 1:378–409
54. Calabrese EJ, Staudenmayer JW, Stanek EJ, Hoffmann GR (2006) Hormesis outperforms threshold model in National Cancer Institute antitumor drug screening database. Toxicol Sci 94:368–78
55. Calabrese EJ, Baldwin LA (2000) Radiation hormesis: Its historical foundations as a biological assumption. Hum Exp Toxicol 19:41–75
56. Calabrese EJ, Baldwin LA (2003) The hormetic dose-response model is more common than the threshold model in toxicology. Toxicol Sci 71:246–50
57. Calabrese EJ (2005) Paradigm lost, paradigm found: the re-emergence of hormesis as a fundamental dose response model in the toxicological sciences. Environ Pollut 138:378–411
58. Calabrese EJ (2002) Hormesis: changing view of the dose response: a personal account of the history and current status. Mutat Res 511:181–9
59. Calabrese EJ, Baldwin LA (2001) Hormesis: a generalizable and unifying assumption. Crit Rev Toxicol 31:353–424
60. Calabrese EJ, Baldwin LA (2001) The frequency of U-shaped dose responses in the toxicological literature. Toxicol Sci 62:330–33
61. Southam CM, Erhlich J (1943) Effects of extracts of Western red-cedar heartwood on certain wood-decaying fungi in culture. Phytopathology 33:517–24
62. Sanders CL (2006) Hormesis as a confounding factor in epidemiological studies of radiation carcinogenesis. Korean Assoc Radiat Prot 31:69–89
63. Scott BR (2008) Low-dose-radiation stimulated natural chemical and biological protection against lung cancer. Dose Respnse 6:299–318
64. Olivieri G, Bodycote J, Wolff S (1984) Adaptive response of human lymphocytes to low concentrations of radioactive thymidine. Science 23:594–7
65. Azzam EI, Raaphorst GP, Mitchell RE (1994) Radiation-induced adaptive response for protection against micronucleus formation and neoplastic transformation in C3H 10T1/2 mouse embryo cells. Radiat Res 138:S28–S31
66. Day TK, Zheng G, Hooker AM et al (2006) Extremely low priming doses of X radiation induced adaptive response for chromosomal inversions in pKZ1 mouse prostate. Radiat Res 166:757–66
67. Day TK, Zheng G, Hooker AM et al (2007) Adaptive response for chromosomal inversions in pKZ1 mouse prostate induced by low doses of X radiation delivered after a high dose. Radiat Res 167:682–92
68. Azzam EI, de Toledo SM, Raaphorst GP, Mitchel RE (1996) Low-dose ionizing radiation decreases the frequency of neoplastic transformation to a level below the spontaneous rate in C3H 10T1/2 cells. Radiat Res 146:369–73
69. Redpath JL, Liang D, Taylor TH et al (2001) The shape of the dose-response curve for radiation-induced neoplastic transformation in vitro: evidence for an adaptive response against neoplastic transformation at low doses of low-LET radiation. Radiat Res 156:700–7
70. Liu S-Z (2007) Cancer control related to stimulation of immunity by low-dose radiation. Dose Response 5:39–47
71. Ina Y, Sakai K (2005) Activation of immunological network by chronic low-dose-rate irradiation in wild-type mouse strains: analysis of immune cell populations and surface molecules. Int J Radiat Biol 81:721–9
72. Sakai K, Hoshi Y, Nomura Y et al (2003) Suppression of carcinogenic process in mice by chronic low dose rate gamma-irradiation. Int J low Radiat 1:142–6
73. Sakamoto K, Myojin M, Hosoi Y et al (1997) Fundamental and clinical studies on cancer control with total or upper-half body irradiation. J Jpn Soc Ther Radiol Oncol 9:161–75

74. Sakamoto K (2004) Radiobiological basis for cancer therapy by total or upper-half body irradiation. Nonlin Biol Toxicol Med 2:293–316
75. Sakai K, Nomura T, Ina Y (2006) Enhancement of bio-protective functions by low dose/dose-rate radiation. Dose Response 4:327–32
76. Johnson A (2008) Publish and be wrong. The Economist 9th Oct 2008
77. Sanders CL, Sohn S (2006) Evidence for radiation hormesis from epidemiological studies of cancer. World J Nucl Med 5(Suppl 1): S258–60
78. Ioannidis JP (2005) Why most published research findings are false. PLOS Med 2(8): e124. http://medicine.plosjournals.org/perlserv?request = get-document&doi = 10.1371/journal.pm
79. Luckey TD (2007) Radiation prevents much cancer. Int J Low Radiat 4:336–44
80. Cohen BL (1995) Test of the linear no-threshold theory of radiation carcinogenesis for inhaled radon decay products. Health Phys 68:157–74
81. Atkinson WD, Law DV, Bromley KJ et al (2004) Mortality of employees of the United Kingdom Atomic Energy Authority, 1946–97. Occup Environ Med 61:577–85
82. Ivanov V, Iiyin L, Gorski A et al (2004) Radiation and epidemiological analysis for solid cancer incidence among nuclear workers who participated in recovery operations following the accident at the Chernobyl NPP. J Radiat Res (Tokyo) 45:41–4
83. Hall P, Mattsson A, Boice JD (1996) Thyroid cancer after diagnostic administration of iodine-131. Radiat Res 145:86–92
84. Holm LE, Wiklud K, Lundell G et al (1988) Thyroid cancer after diagnostic doses of iodine-131: a retrospective cohort study. J Natl Cancer Inst 80:1133–8
85. Holm LE, Hall P, Wiklud K et al (1991) Cancer risk after iodine-131 therapy for hyperthyroidism. J Natl Cancer Inst 83:1072–7
86. Thompson RE, Nelson DF, Popkin JH, Popkin Z (2008) Case–control study of lung cancer risk from residential radon exposure in Worcester County, Massachusetts. Health Phys 94:228–41
87. Luckey TD (2008) Abundant health from radioactive waste. Int J Low Radiat 5:71–82
88. UNSCEAR (2000) 2000 report to the general assembly, with annexes. Annex J, vol II. United Nations, New York, pp 451–566
89. Schull WJ, Otakee M, Neel JV (1981) Genetic effects of the atomic bomb: a reappraisal. Science 213:1220–5
90. Neel JV, Schull WJ, Awa AA et al (1990) The children of parents exposed to atomic bombs. Estimates of the genetic doubling dose of radiation for humans. Am J Hum Genet 46: 1053–79
91. Mine M, Okumura Y, Ichimaru M et al (1990) Apparently beneficial effect of low to intermediate doses of A-bomb radiation on human lifespan. Int J Radiat Biol 58:1035–43
92. Stewart AM (1990) Healthy worker and healthy survivor effects in relation to the cancer risks of radiation workers. Am J Ind Med 17:151–4
93. Li C-Y, Sung F-C (1999) A review of the healthy worker effect in occupational epidemiology. Occup Med 49:225–9
94. Monson RR (1986) Observations on the healthy worker effect. J Occup Med 28:425–33
95. Matanoski GM (1991) Health effects of low-level radiation in shipyard workers. Final Report. Report No. DOE DE-AC02–79EV10095. U.S. Department of Energy, Washington, DC
96. Sponsler R, Cameron JR (2005) Nuclear shipyard worker study (1980–1988): a large cohort exposed to low-dose-rate gamma radiation. Int J Low Radiat 1:463–78
97. Thierry-Chef I et al (2007) The 15-country collaborative study of cancer risk among radiation workers in the nuclear industry: study of errors in dosimetry. Radiat Res 167:380–95
98. Cardis E, Vrijheid M, Blettner M et al (2007) The 15-country collaborative study of cancer risk among radiation workers in the nuclear industry: estimates of radiation-related cancer risks. Radiat Res 167:396–416
99. Luan YC, Shieh MC, Chen HT et al (2006) Re-examining the health effects of radiation and its protection. Int J Low Radiat 3:27–44

100. Upton AC (2001) Radiation hormesis: data and interpretations. Crit Rev Toxicol 31:681–95
101. Choi NC, Timothy AR, Kaufman SD et al (1979) Low dose fractionated whole body irradiation in the treatment of advanced non-Hodgkin's lymphoma. Cancer 43:1636–42
102. Taylor LS (1980) Some non-scientific influences of radiation protection standards and practice. In: Fifth International Congress of the International Radiation Protection Association, vol 1. The Israel Health Physics Society, Jerusalem, pp 307–19
103. Tubiana M, A Aurengo, D Averbeck et al (2006). The debate on the use of linear no threshold for assessing the effects of low doses. J Radiol Prot 26:317–24
104. Cuttler JM (2007) Health effects of low level radiation: when will we acknowledge the reality? Dose Response 5L 292–8
105. Maynard K (2008) Special issue on hormesis. Am J Pharm Toxicol 3:1–3

# Index

MIX
Papier aus verantwortungsvollen Quellen
Paper from responsible sources
FSC® C105338

If you have any concerns about our products,
you can contact us on
**ProductSafety@springernature.com**

In case Publisher is established outside the EU,
the EU authorized representative is:
**Springer Nature Customer Service Center GmbH**
**Europaplatz 3, 69115 Heidelberg, Germany**

Printed by Libri Plureos GmbH
in Hamburg, Germany